W0037676

Lecture Notes in Physics

Lecture Notes
in Physics

Edited by H. Araki, Kyoto, J. Ehlers, München, K. Hepp, Zürich
R. Kippenhahn, München, H. A. Weidenmüller, Heidelberg
and J. Zittartz, Köln

183

J. D. Gunton
M. Droz

Introduction to the Theory of Metastable and Unstable States

Springer-Verlag
Berlin Heidelberg GmbH 1983

Authors

J. D. Gunton
Department of Physics, Temple University
Philadelphia, PA 19122, USA

M. Droz
Département de Physique Théorique, Université de Genève
1211 Genève 4, Switzerland

ISBN 978-3-540-12306-4   ISBN 978-3-540-39894-3 (eBook)
DOI 10.1007/978-3-540-39894-3

Originally published by Springer-Verlag Berlin Heidelberg New York Tokyo in 1983

2153/3140-543210

PREFACE

This monograph is an introduction to the theory of the dynamics of metastable and unstable states. This is an active branch of research in which many fundamental questions remain unanswered. Therefore our discussion of the theory is by necessity incomplete. We do attempt, however, to discuss the basic ideas so far formulated in order that the reader can have an overall view of the current state of the theory. We do not give an exhaustive list of references to the theoretical and experimental work in this field, since recent reviews on this subject (which are mentioned in the text) are reasonably comprehensive. We do, however, provide a list of selected references for further study at the end of each chapter.

This monograph has been developed from a lecture series given in the "Troisième Cycle de la Physique en Suisse Romande" (CICP) by one of us (JDG).

We wish to acknowledge the financial support of the CICP, of the Swiss National Science Foundation and of a grant from the US.NSF (DMR -8013700).

Geneva, March 1983   J.D. Gunton

                                      M. Droz

TABLE OF CONTENTS

CHAPTER 1. INTRODUCTION

This monograph deals with the theory of metastable and unstable states. Metastable states of matter are well known phenomena of nature. A familiar example is a supercooled vapor which with proper precautions can be maintained in a metastable state for quite a long time. Eventually, however, this vapor will condense into the more stable, equilibrium liquid phase. This condensation requires the occurrence of thermal fluctuations of density of a certain critical size (the "critical droplet"). A certain activation energy is required for the formation of such droplets.

Liquid droplets that form which are larger than this critical size will grow with time, while smaller droplets will shrink. The rate of birth of these droplets involves the theory of homogeneous nucleation, which has been a subject of research for at least fifty years. Although the basic ideas of nucleation theory are rather well known, a completely satisfactory first principles theory is not yet available. (Nucleation theory however is much better developed than the theory of unstable states which we come to later.)

One aspect of metastability which is particularly striking is just how "stable" certain metastable states can be. That is, the lifetime for the "decay" of a metastable state can be extraordinarily large. The classic example is, of course, diamond, which is a metastable form of carbon. Another example of how "stubbornly resistant" an ordinary material can be to a change of its phase was given by Cahn [1], in an article which provides a particularly clear introduction to the subject of metastability and instability. He considered the case of superheated liquid water at 110°C and atmospheric pressure. (Recall that water should boil and change to vapor at 100°C at atmospheric pressure.) He points out that if one filled the volume of the universe with this superheated water, it would remain metastable for as long as 10^{10} years, which is of the order of the age of the universe. Thus water would not boil at 110°C unless nucleation catalysts (such as impurities) were present.

Since a proper understanding of metastability requires a dynamical description, a theory of the time evolution of a phase separating system is necessary. Except under special circumstances ("late stage" growth, where one can use the Lifshitz-Slyozov theory), a satisfactory dynamical theory has not been developed. The most interesting work to date in this direction is due to Langer and Schwartz [2] who attempted to solve the very difficult, nonlinear dynamical equations of motion necessary to describe a phase separating system under circumstances in which the birth and growth of droplets are simultaneously occurring.

We will return to other examples of metastable states later in this chapter. In all such cases, however, there is an analogue of the "critical droplet", mentioned above in the supercooled vapor, which is a crucial ingredient in homogeneous nucleation theory.

We now briefly describe the problem of unstable states. This is a situation which has received considerable attention in metallurgy, but until relatively recently has received rather little attention in other scientific disciplines. To describe a typical situation in which a system can be prepared in a thermodynamically unstable state, we refer to Fig. 1.1. Here a phase diagram is shown for a system such as a binary alloy or a binary fluid, in which the order parameter is the average concentration c of one of the two species. As is well known [3], below a certain temperature, T_c , thermodynamic equilibrium corresponds to a coexistence of two phases, one rich in one of the species, the other poor in this species. This equilibrium situation is indicated in Fig. 1.1 by a solid line known as the coexistence curve. The dashed curve shown in this figure is the mean field prediction for the "spinodal" curve. This concept arises from any mean field theory, such as the van der Waals theory, in which one has a "van der Waals loop", as shown in Fig. 1.2. Here a chemical potential μ is plotted versus the concentration of one of the two species for a fixed temperature $T < T_c$ It is well known that the stable equilibrium state corresponds to the solid curve, so that a system whose concentration is between c_A and c_B has a thermal equilibrium state which is a mixture of the two phases (with concentrations c_A and c_B) given by the well-known lever rule. What is of interest in Fig. 1.2, however, is the van der Waals loop, which in a classical theory provides a definition of metastable and unstable states. States for which $(\partial\mu/\partial c)_T \gtrless 0$ are termed metastable and unstable, respectively and are, of course, nonequilibrium states.

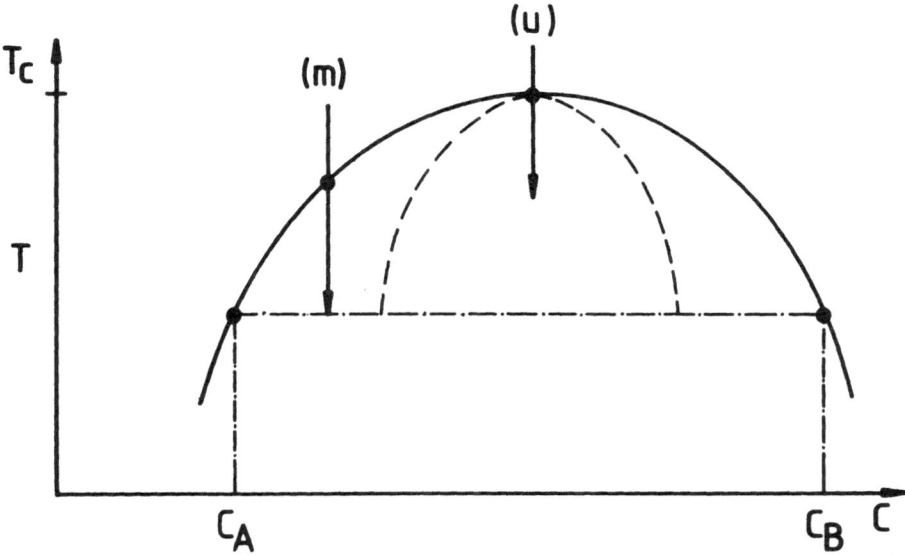

<u>Figure 1.1</u> : The coexistence curve (solid line) and classical spinodal curve (dashed line) are shown schematically for a system such as a binary fluid or binary alloy. Typical quenches into the metastable (m) and the unstable (u) regions are also shown.

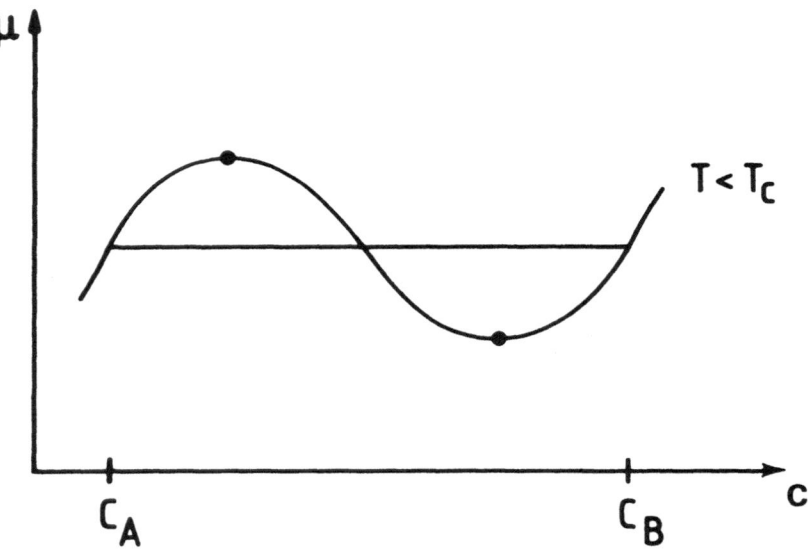

<u>Figure 1.2</u> : The chemical potential μ as a function of concentration
c for T < T_c , as predicted by mean field theory. The
horizontal line indicates the equilibrium chemical potential, while the van der Waals loop indicates the regions
of metastability and instability. The dots indicate the
classical spinodal points, for which $(\partial\mu/\partial c)_T = 0$.

The two points at which $(\partial\mu/\partial c)_T = 0$ are termed spinodal points. The locus of such points as a function of temperature is the spinodal curve shown in Fig. 1.1.

Consider now an experiment in which a system such as a binary alloy is "quenched" rapidly from an initial one-phase state $(T > T_c)$ to a temperature $T < T_c$ at a constant concentration (such as the critical concentration c_c) which brings the system into the unstable region of the phase diagram; that is, to a point (c,T) for which $\partial\mu/\partial c < 0$ in the classical theory. The path u in Fig. 1.1 describes such a process taking place at the critical concentration c_c. The time evolution of the system following such a quench involves the dynamics of an unstable state. In contrast to a quench to a metastable state (path m in Fig. 1.1) no activation energy is required for the system to begin to phase separate following the quench u. Also, in contrast to a metastable state in which a localized droplet of finite size is required to initiate phase separation, the initial evolution of an unstable state involves infinitesimally small long-wave fluctuations of the local order parameter (such as the local concentration). This initial process is known as spinodal decomposition and is physically manifest as a finely dispersed precipitate which gradually coarsens.

An excellent example showing the respective decays of a metastable and unstable state is shown in Fig. 1.3. This is a transmission electron microscope study [4] of the alloy FeAl. The left and right hand sets of photographs show the evolution of a metastable system phase separating via the birth and growth of droplets. The center photographs show the spinodal decomposition and subsequent coarsening of an unstable state of the system. Such alloys provide excellent examples for the study of metastable and unstable states, due in part to the small diffusion constants involved which results in very long time scales for phase separation. It should also be noted that the interconnectivity shown in Fig. 1.3 provides an interesting example of pattern formation in an equilibrating system.

Several other remarks are worth making with respect to Fig. 1.3. First of all, the problem of metastable and unstable states involves systems which are far from equilibrium. Even after 10'000 minutes in this particular case, the system is far from a thermal equilibrium state

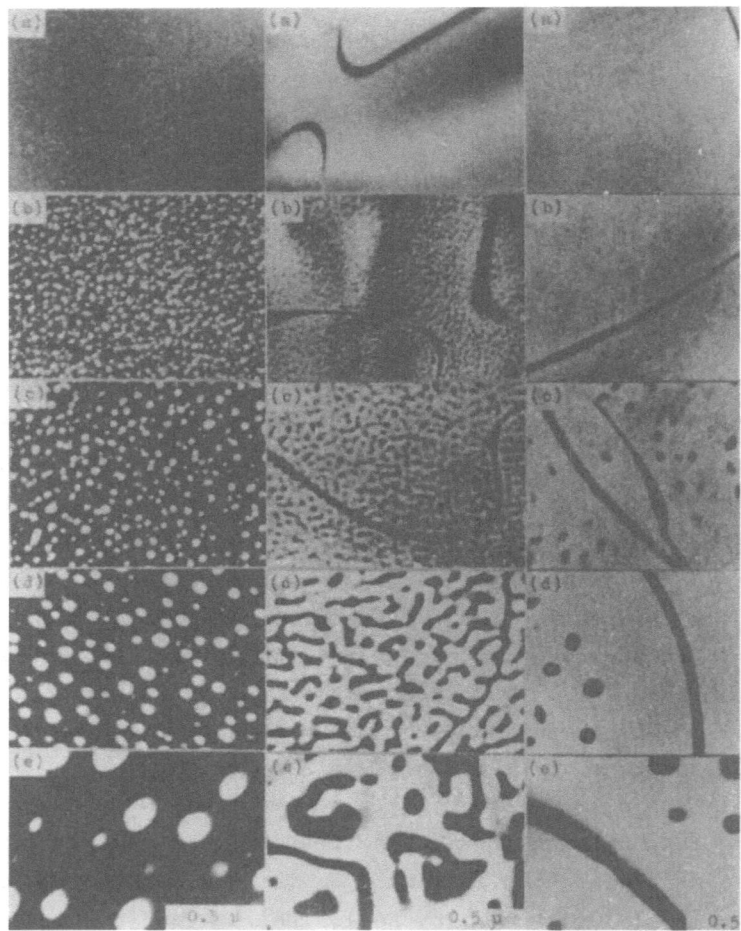

<u>Figure 1.3</u> : Domain structures imaged with B_2 superlattice reflection in 23.0, 24.7 and 24.9 at. % Al alloys, from left to right. The samples are quenched from 630°C and annealed at 570°C in the case of 23.0 and 24.7 at. % Al alloys and at. 568°C in the case of 24.9 at. % Al alloy.
a) as quenched; b) annealed for 15 min. in 23.0 and 24.9 at. % Al and for 10 min. in 24.7 at. % Al alloy; c) 100 min.; d) 1000 min.; e) 10000 min..
(From Oki, Sagana and Eguchi, Ref. [4].)

of bulk two phase coexistence. Secondly, after an initial time such that well defined interfaces are formed, the study of the time evolution of metastable and unstable states involves the dynamics of random interfaces. Indeed, from this point of view certain aspects of the dynamics of metastable and unstable states provide a striking example of the general problem of the dynamics of topological singularities. Other examples include vortices in superfluid He^4 and dislocations. Thirdly, if one examines Fig. 1.3 one can see that to a first approximation there is a self-similarity or dynamical scaling involved in the evolution of the phase separating system. That is, one can imagine reducing the bottom photographs in each set by some suitable scale factor such that each of these looks very much like the corresponding photograph immediately above it. Thus, to a first approximation the pattern formation is invariant (in a loose sense) under a time-dependent length rescaling. These three observations will in fact be the subject of several chapters of this monograph.

It is also worth noting what one would see in a small angle scattering experiment of neutrons (or X-rays, light, ect, depending on the particular system of interest). A typical experimental result [5] for a glass is shown in Fig. 1.4.

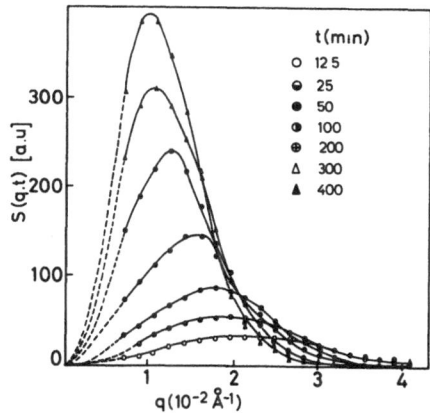

Figure 1.4 : Plot of the scattering intensity $S(q,t)$ as a function of the scattering wave number q for different times for the system $B_2O_3 - PbO - Al_2O_3$ with composition 80-15-5 (wt.%). (From A. Graievich and J.M. Sanchez, Ref. [5].)

One sees that a peak develops in the scattering intensity $S(q,t)$ as the phase separation develops. The peak intensity increases and the peak position $q_m(t)$ decreases as time increases. One usually assumes that $q_m^{-1}(t)$ is proportional to the characteristic length scale of the system, so that as the pattern coarsens it is reasonable that $q_m(t)$ decreases. Since precision studies of $S(q,t)$ are possible for many systems, a theory for this scattering intensity would be most useful. A discussion of the existing theory is given in later chapters. We also show in Fig. 1.5 an example of a dynamical scaling of the structure factor for the system shown in Fig. 1.4.

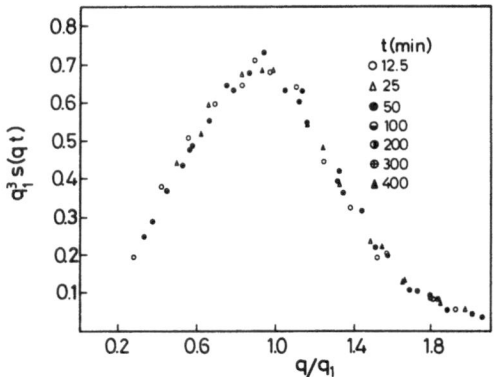

Figure 1.5 : Plot of $q_1^3 S(q,t)$ as a function of q/q_1 where q_1 is the first moment of the scattering intensity $S(q,t)$ for the system shown in Fig. 1.4. This plot demonstrates the dynamical scaling behaviour discussed in the text.

Dynamical scaling asserts that $q_1^3(t) S(q,t) = F(q/q_1, t)$, to a good first approximation, where $q_1(t)$ is some characteristic time dependent wave number. In Fig. 1.5 $q_1(t)$ is the first moment of $S(q,t)$. Such a scaling behaviour has now been observed experimentally for a variety of different systems and is the subject of considerable theoretical investigation.

Before briefly outlining the contents of this monograph, we pause to note the enormous diversity of systems in which metastable and unstable states have been studied, either experimentally or theoretically. These include

- simple fluids (gas-liquid transitions);
- binary fluids;
- binary alloys;
- superfluids and superconductors;
- physisorption and chemisorption systems;
- intercalation compounds;
- polymer blends;
- gels;
- lasers;
- electron-hole condensation in semiconductors;
- geological systems (minerals);
- chemically reacting systems;
- metals;
- glasses and crystalline ceramics;
- order-disorder systems;
- coherent hydrogen-metal systems;
- magnetic systems;
- astrophysics.

It is clearly impossible in a few chapters to discuss all of these systems. For pedagogical reasons we have chosen to discuss primarily a simple model of a binary alloy which illustrates most of the basic theoretical ideas involved in systems whose order parameter is conserved. We also discuss in certain chapters a model in which the order parameter is not conserved (such as a simple antiferromagnet), since a crucial distinction in the evolution of a system depends on whether or not the order parameter is locally conserved. References to theoretical and experimental work dealing with all the above topics (except astrophysics see [6]) can be found in a recent review article by Gunton, San Miguel and Sahni [7].

The outline of this monograph is the following. In Chapter 2 we discuss a simple Ising model of a binary alloy. We also introduce the concept of a coarse grained Helmholtz free energy functional, which is important for a discussion of metastable and unstable states. We also summarize some recent work related to the calculation of this functional.

In Chapter 3 we describe a continuum dynamical model of a binary alloy, suitable for a discussion of metastable and unstable states. We discuss the master equation and closely related Fokker-Planck equation for the binary alloy. We also discuss an equivalent nonlinear Langevin equation for this model and introduce a nonlinear Langevin equation for a model in which the order parameter is not conserved. We also derive equations for non equilibrium correlation functions related to the scattering intensity mentioned above. In Chapter 4 we discuss the classical theory of nucleation (which is based on a classical droplet model), which provided the foundation for all subsequent theoretical work in homogeneous nucleation theory. In Chapter 5 we discuss a continuum generalization of the classical droplet model and develop a so-called "drumhead model". We then use this to obtain the analytic continuation of the free energy of a stable phase to a metastable phase. In particular we describe the imaginary part of the free energy of this analytic continuation, since the nucleation rate is proportional to this imaginary part. This fact is shown in a field theory calculation of the nucleation rate discussed in Chapter 6. We believe that this field theory formalism, developed by Langer [7], provides the basis for a systematic theory of nucleation, although a completely convincing experimental confirmation of this (or any other extension of classical nucleation theory) is not yet available. In Chapter 7 we turn to a theoretical description of the dynamics of unstable states. There is no satisfactory theory in this field. We therefore discuss the important linear theory which at least qualitatively describes the very early stages of spinodal decomposition. We then describe in some detail the most satisfactory attempt to date to deal with the important nonlinearity in the equation of motion for the structure factor. (This nonequilibrium correlation function is proportional to the scattering intensity in the absence of multiple scattering effects.) We also summarize some of the inadequacies of the existing theory, in order to possibly provide the basis for future improvement in this important field. In Chapter 8 we describe the Lifshitz-Slyozov late stage growth theory. This is probably the only "exact" dynamical theory available which treats nonlinear effects in the kinematics of first order phase transitions. In Chapter 9, we discuss a kinetic drumhead model, which seems to provide an important starting point for a theory of the dynamics of random interfaces, at least under circumstances in which the interfaces are rather gently curved and rather thin. In Chapter 10 we summarize a recent application of this kinetic drumhead model to the case of a simple antiferromagnet. In this case an explicit expression for the nonequilibrium structure factor for a nonconserved order parameter is obtained which exhibits dynamical scaling. The scaling function agrees reasonably well with existing computer simulation data for the two and three dimensional antiferromagnet.

REFERENCES - Chapter 1.

[1] J.W. Cahn, in "Critical Phenomena in Alloys, Magnets, and Superconductors", McGraw-Hill, edited by R.E. Mills, E. Ascher and
R.J. Jaffee (1971).

[2] J.S. Langer and A.J. Schwartz, Phys. Rev. A21, 948 (1980).

[3] H.E. Stanley, "Introduction to Phase Transitions and Critical Phenomena", Ed. Clarendon Press Oxford (1971).

[4] K. Oki, H. Sagana and T. Eguchi, J. de Physique C7, 414 (1977).

[5] A. Craievich and J.M. Sanchez, Phys. Rev. Lett. 47, 1308 (1981).

[6] A. Guth, Phys. Rev. D23, 347 (1981).
A. Albrecht and P.J. Steinhardt, Phys. Rev. Lett. 48, 1220 (1982).

[7] J.D. Gunton, M. San Miguel and P.S. Sahni, to be published in Vol.
8, "Phase Transitions and Critical Phenomena", Academic Press
(London), edited by C. Domb and J.L. Lebowitz (1983).

GENERAL REFERENCES

[G1] F.F. Abraham, "Homogeneous Nucleation Theory", Academic Press
(New York and London), (1974). This is one of the standard texts
on homogeneous nucleation theory.

[G2] V.P. Skripov, "Metastable Liquids", J. Wiley and Sons Inc.,
(New York), (1974). Another useful text on metastability.

[G3] V.P. Skripov and A.V. Skripov, Sov. Phys. Uspkeki 22, 389 (1979).
This is a good recent review of the subject of spinodal decomposition.

[G4] Reference [7] above (J.D. Gunton, M. San Miguel and P.S. Sahni)
contains an extensive review of both the theoretical and experimental aspects of metastable and unstable states. In particular,

it contains a list of references for each of the chapters discussed
in this monograph.

[G5] O. Penrose and J.L. Lebowitz, in "<u>Studies in Statistical Mechanics</u>",
Vol. 7, North-Holland (Amsterdam), edited by J.L. Lebowitz and
E. Montroll.

CHAPTER 2.   A SIMPLE MODEL OF BINARY ALLOYS

The goal of this chapter is to define a model describing binary alloys (or binary fluids) and to study its static properties. This is the well-known Ising model which we present here for completeness in Section 2.1. In Section 2.2, we discuss the concept of a coarse-grained free energy functional. Finally in Section 2.3 we review some recent calculation of this free energy functional by Monte Carlo and renormalization group methods.

2.1 Ising model for binary alloys

A binary alloy is constituted by two kinds of components A and B sitting at the sites of a regular d dimensional lattice. Each site $\vec{x}_i$ of the lattice can be occupied either by the atomic species A or B. It is then suitable to introduce the variable c_i defined by

$$c_i = \begin{cases} +1 & \text{if the site } \vec{x}_i \text{ is occupied by a constituent of type } A \\ \\ -1 & \text{if the site } \vec{x}_i \text{ is occupied by a constituent of type } B \end{cases}$$

If N is the number of sites of the lattice, then

$$c = \frac{1}{N} \sum_i c_i = \frac{N_A - N_B}{N} \tag{2.1}$$

is the concentration of the alloy. N_A and N_B are respectively the number of particles (atoms or molecules) of the A or B components. Indeed, since $N = N_A + N_B$ we have

$$c_A = \frac{1}{2}(1 + c)$$

$$\tag{2.2}$$

$$c_B = \frac{1}{2}(1 - c)$$

where $c_{A,B} = \frac{N_{A,B}}{N}$ are the concentrations of the components A and B. Let us suppose that the particles interact only if they are nearest neighbours on the lattice. This is a reasonable approximation for particles interacting via short range forces. However, more realistic models of alloys must include other types of interactions. (See, for example, van Baal [1].) For nearest neighbour particles of type α, β, (α, β being A or B) the interaction is $\varepsilon_{\alpha\beta}$. If the configuration has N_{AA} pairs of nearest neighbours of type A, N_{BB} of type B and N_{AB} of type A,B (or B,A), the energy of this configuration is :

$$E = \varepsilon_{AA} N_{AA} + \varepsilon_{BB} N_{BB} + \varepsilon_{AB} N_{AB} \ . \tag{2.3}$$

This energy can be easily expressed in terms of the variables c_i. Using the Knonecker function :

$$\delta_x = \left\{ \begin{array}{ll} 1 & \text{if } x = 0 \\ \\ 0 & \text{otherwise} \end{array} \right. \tag{2.4}$$

(2.3) can be rewritten as :

$$E = \sum_{\langle i,j \rangle}^{*} \{ \varepsilon_{AA} \delta_{c_i - 1} \delta_{c_j - 1} + \varepsilon_{BB} \delta_{c_i + 1} \delta_{c_j + 1}$$

$$+ \varepsilon_{AB} (\delta_{c_i - 1} \delta_{c_j + 1} + \delta_{c_i + 1} \delta_{c_j - 1}) \} \ , \tag{2.5}$$

where $\sum_{\langle ij \rangle}$ means summation over the lattice sites i and all the j nearest neighbours of i. Moreover, the asterisk means that, for this configuration, the c_i are constrained by the condition

$$\sum_i c_i = Nc \tag{2.6}$$

where c is the concentration associated with this particular configuration. Thus the summation (2.5) is restricted by the constraint (2.6). Note that due to the fact that c_i is plus or minus one, we have the identity :

$$\delta_{c_i \pm 1} = \mp \frac{1}{2} [c_i \mp 1] \ . \tag{2.7}$$

Thus (2.5) can be rewritten in terms of the c_i themselves. The energy of any configuration of the binary alloy associated with a concentration c can be written in the form (2.5). Accordingly one can describe the binary alloy by a Hamiltonian $\mathcal{H}$ whose form is (using (2.7) and

and (2.5)) :

$$\mathcal{H} = E_o - J \sum_{\langle ij \rangle}^{*} c_i c_j - h \sum_i^{*} c_i \tag{2.8}$$

with

$$E_o = \frac{Nz}{4} [\varepsilon_{AA} + \varepsilon_{BB} + 2\varepsilon_{AB}] \tag{2.9}$$

$$J = -\frac{1}{4} [\varepsilon_{AA} + \varepsilon_{BB} - 2\varepsilon_{AB}] \tag{2.10}$$

$$h = \frac{z}{2} [\varepsilon_{BB} - \varepsilon_{AA}] \tag{2.11}$$

z is the coordination number of the lattice, i.e. the number of nearest neighbours of a given site. Note that depending on the value of ε_{AA}, ε_{BB} and ε_{AB}, J and h may be positive or negative.

Given the model defined by the Hamiltonian (2.8), we have now to determine its equilibrium properties using standard statistical mechanics. Due to the constraint (2.6) on the Hamiltonian it is preferable to work within the grand canonical ensemble, in which the concentration c of the alloy is not fixed. In terms of the fugacity $\mathfrak{z}$, related to the chemical potential μ through the relation

$$\mathfrak{z} = \exp \beta\mu, \quad \beta = 1/k_B T \tag{2.12}$$

k_B being the Boltzmann constant and T the temperature, the grand canonical partition function is

$$Z_{GC}(\mu,N,T) = \sum_{N_A=0}^{N} \mathfrak{z}^{N_A} \operatorname*{Tr}_{(c_i)}^{(N_A)} \exp - \beta\mathcal{H} \tag{2.13}$$

where $\operatorname*{Tr}_{(c_i)}^{(N_A)}$ means summation over the possible configurations compatible with a given number of constituent of type A. Using (2.12) and (2.6) we can finally rewrite (2.13) as

$$Z_{GC}(\mu,N,T) = \operatorname*{Tr}_{\{c_i\}} \exp - \beta\bar{\mathcal{H}} \tag{2.14}$$

where the summation is over all the possible configurations and

$$\bar{H} = E_o - J \sum_{<ij>} c_i c_j - \bar{\mu} \sum_i c_i \qquad\qquad (2.15)$$

with $\bar{\mu} = \mu + h$.

Knowing the grand canonical partition function, we can compute the thermodynamic quantities of interest, and derive the phase diagram [2]. Then a given binary alloy, i.e., an alloy with a given initial concentration c is represented by an iso-concentration line in the phase diagram given by (2.14).

Remarks : I. The above derivation has been performed for a binary alloy. However, a similar analysis can be done for a binary fluid. Although one does not have a lattice for the fluid case, one can divide the fluid into small cells of volume v_o. v_o is chosen in such a way that only one particle (A or B) can be in any cell. If the center of gravity of particle is in the cell $\vec{x}_i$, the cell is said to be occupied by this particle. As for the binary alloy, a variable c_i can be then associated with the cell $\vec{x}_i$. All that has been said for the binary alloy remains true for the binary fluid.

II. In magnetism a particular model plays an important role due to its relative simplicity on one hand and the richness of its physical properties on the other hand. This is the Ising model [2]. In this model, the sites of a lattice are occupied by magnetic moments or spins, s_i, which can assume only two states "up" or "down". Thus $s_i = \pm 1$. Each spin s_i interacts with its nearest neighbours via an exchange interaction J_I and may interact with an external magnetic field h_I. The Ising Hamiltonian reads then :

$$H_I = E_{0I} - J_I \sum_{<ij>} s_i s_j - h_I \sum_i s_i . \qquad\qquad (2.16)$$

It is easy to show that the canonical partition function of the Ising model is equal to the grand canonical partition function of our binary alloy, given a proper identification of the parameters [2]. The chemical potential μ is related to the magnetic field h_I. Working with a fixed concentration c is equivalent to working at fixed magnetization in the Ising case. Thus, via a proper transcription of the parameters, one may use the vast amount of results derived for the Ising model to study the binary alloy (or the lattice gas model of binary fluids). In particular, the phase diagram of the binary alloy will be similar to that of the Ising model in a field.

Phase diagram and equilibrium free energy

In order to find the phase diagram for our binary alloy we have to compute the grand canonical function (2.14) and derive the thermodynamic quantities of interest. It is well known that, except for special cases (two dimensional lattices with zero field or one dimensional chains), one cannot compute Z_{GC} exactly. One has to use some approximate methods. The usual approximations are the Bragg-Williams [3], mean-field [4], Monte Carlo [5], series expansions [6] or renormalization group approximations [7]. In principle renormalization group methods provide the most powerful theoretical tools for such studies, although in certain cases Monte Carlo or phenomenological scaling calculations are easier to perform. In general, mean field and Bragg-Williams theories provide only qualitatively useful results for systems with short range forces, but give exact results for systems with long range forces [8]. A vast literature is devoted to these questions. One now has an accurate knowledge of many models of phase transitions, including the Ising model. We are not going to derive explicitly the phase diagram for the binary alloy here. Let us just recall the main qualitative features. The situation can be summarized by projecting the phase diagram on the planes $(\bar{\mu};c)$, $(\bar{\mu};T)$ and $(T;c)$, as shown on the Figs 2.1, 2.2 and 2.3.

Let us consider a given binary alloy with concentration c at a sufficiently high temperature T_1 above the temperature $T_c(c)$ (see Fig. 2.3). If we lower the temperature to a value T_2 below the temperature $T_c(c)$, the system will decompose into an equilibrium state which consists of the coexistence of two phases. One phase will be rich in the B constituent and have a concentration c'_B, the other rich in the A constituent with concentration c'_A. The proportions of the A and B rich phase are given by the "lever rule", i.e., if $x_{A,B}$ are respectively the proportions of the rich A or B phase :

$$x_A = \frac{c - c'_B}{c'_A - c'_B} \quad \text{and} \quad x_B = \frac{c'_A - c}{c'_A - c'_B}$$

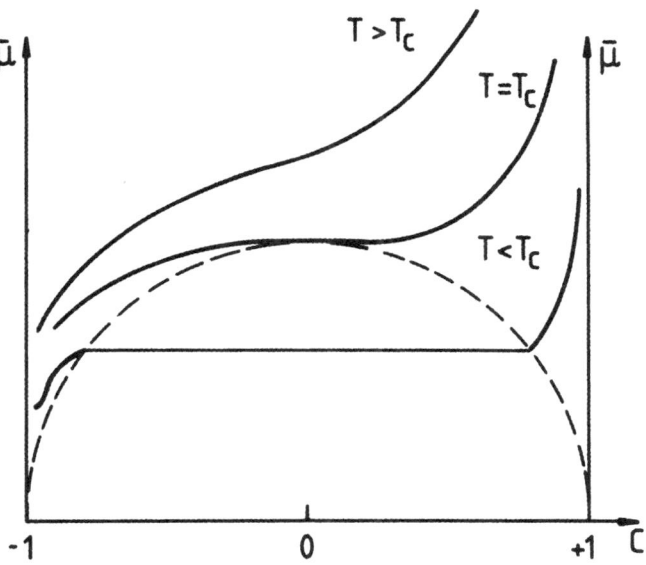

<u>Figure 2.1</u> : Typical isotherms for a simple binary alloy. The dotted
line represents the coexistence curve.

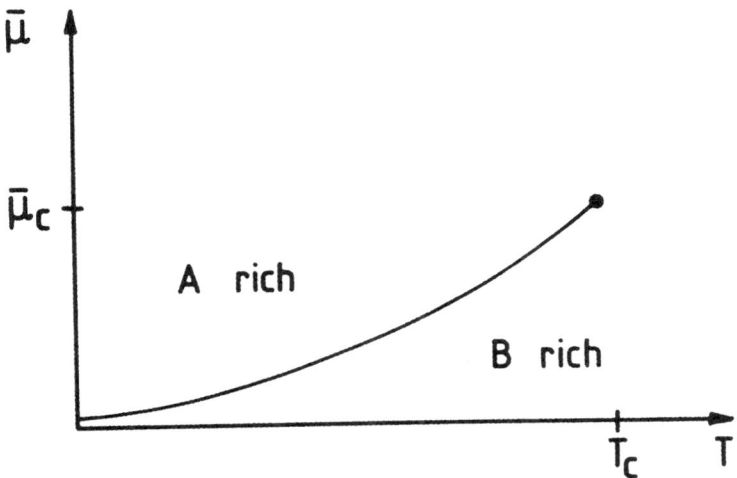

<u>Figure 2.2</u> : The coexistence curve shown in the intensive variables
$(\bar{\mu};T)$. This coexistence curve terminates at a critical
point indicated by a dot.

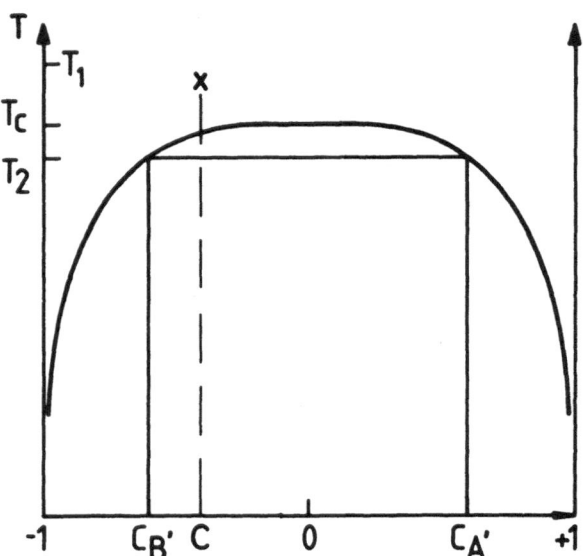

<u>Figure 2.3</u> : Coexistence curve for a simple binary alloy. The alloy with con-
centration c , quenched at temperature T_2 , decomposes into
two phases with concentrations c_A' and c_B'.

The equilibrium free energy density $f(T,c)$ will be a convex function of c for all temperatures. Particularly, for $T = T_2$, $f(T_2,c)$
will have the form sketched on Fig. 2.4.

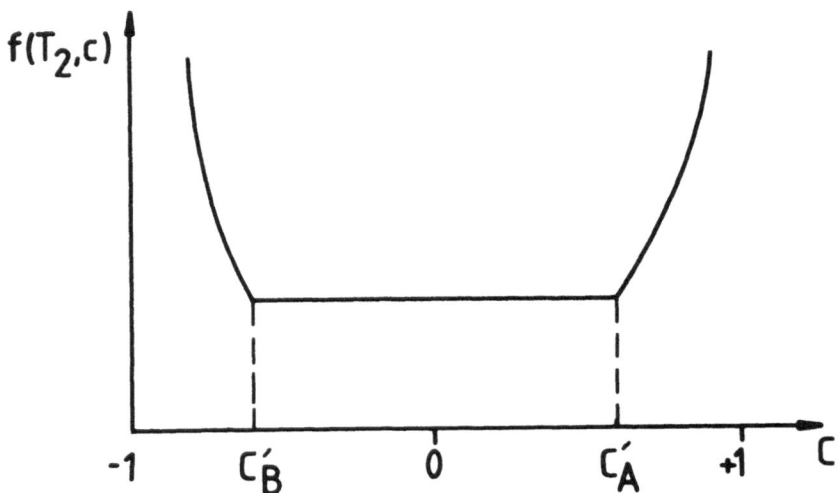

<u>Figure 2.4</u> : The equilibrium free energy density as a function of the
concentration c for a temperature T_2 less than T_c.

According the chemical potential $\tilde{\mu} = \partial f/\partial c$ will have the form sketched on Fig. 1.2.

Thus, the exact equilibrium free energy contains no useful information about metastable or unstable states as defined in Chapter 1. In the following section, we shall introduce a coarse grained free energy in order to get some information about metastable and unstable states for systems with short range forces.

2.2 Coarse-grained free energy functional

The description given in Section 2.1 was purely microscopic. Let us now consider a semi-macroscopic description. Instead of looking at the original lattice (with lattice constant a_o), let us divide our lattice into cubical cells of edge size $L = \ell a_o$, labelled by an index α and centered in $\vec{x}_\alpha$. This is shown in Fig. 2.5 for the two dimensional lattice.

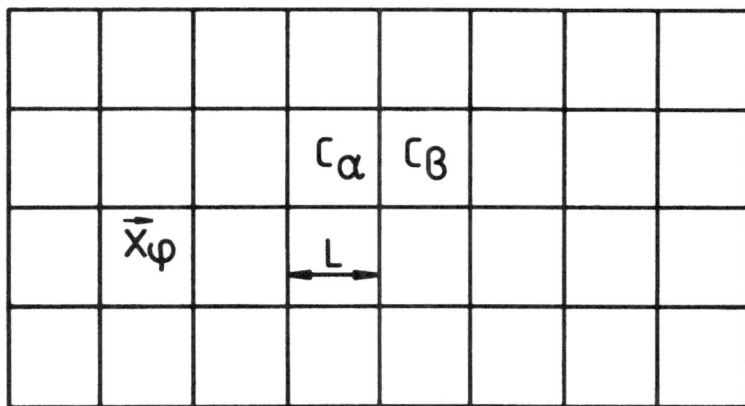

Figure 2.5 : Division of the system into cells of linear dimension L. c_α represents the average value of the order parameter (the average local concentration) in cell α. $\vec{x}_\varphi$ locates the center of cell φ.

The average concentration in the cell α is

$$c_\alpha = \frac{1}{N_\alpha} \sum_{i \in \alpha} c_i \tag{2.17}$$

where N_α is the number of sites in the cell α. Then, the overall concentration is given by

$$c = \frac{1}{M} \sum_{\alpha=1}^{M} c_\alpha \tag{2.18}$$

where M is the number of cells.

Microscopically, the grand canonical partition function is given by (2.14), with the trace involving all the microscopic c_i variables. One can decompose this operation in two steps :

   i) Take the trace over the microscopic configurations $\{c_i\}$ compatible with a given constraint $\{c_\alpha\}$, corresponding to a specific configuration of cell variables.

  ii) Take the trace over all the possible configurations $\{c_\alpha\}$. If $W\{c_\alpha\}$ denotes the number of microscopic configurations $\{c_i\}$ compatible with the constraint $\{c_\alpha\}$, we can write :

$$\begin{aligned} Z_{GC} &= \mathop{Tr}_{\{c_\alpha\}} W\{c_\alpha\} \exp - \beta E\{c_\alpha\} \\ &\equiv \mathop{Tr}_{\{c_\alpha\}} \exp[-\beta F\{c_\alpha\}] \end{aligned} \tag{2.19}$$

where

$$F\{c_\alpha\} = E\{c_\alpha\} - k_B T \log W\{c_\alpha\} . \tag{2.20}$$

F is the sum of an energy and an entropy term and is thus a coarse grained free energy functional. Moreover, instead of working with the cell variables c_α, defined at discretized points $\vec{x}_\alpha$ we can introduce a field $c(\vec{x})$ defined for all $\vec{x}$ and which extrapolates smoothly the $c_\alpha(\vec{x}_\alpha)$. Thus, $c(\vec{x}_\alpha) = c_\alpha(\vec{x}_\alpha)$ varies slowly over distances comparable to the cell size L. Accordingly, the corresponding free energy can be written as an integral over all the space of a free energy density. This density will be composed of a spatially homogeneous part V_L and a spatially inhomogeneous part which describes the slow spatial variations. Thus

$$F_L\{c(\vec{x})\} = \int d\vec{x}\, f_L(c(\vec{x}))$$

$$= \int d^d x\, [\tfrac{1}{2} C_L |\nabla c(\vec{x})|^2 + V_L(c(\vec{x}))] \quad . \tag{2.21}$$

The index L in (2.21) expresses the fact that the coarse-grained free energy thus obtained depends on the size of the cell L used for the average. A similar way to obtain a coarse-grained free energy consists in going to Fourier space and integrating out the short wave length components. This is the strategy proposed by K. Wilson [7] in his original renormalization group approach to critical phenomena.

However, one seldom is able to perform explicitly either of the coarse graining procedures described above starting from a fully microscopic model. One rather assumes a phenomenological Ginzburg-Landau form for the coarse-grained free energy $F\{c(\vec{x})\}$. $F\{c(\vec{x})\}$ has to be such as to reflect the physical properties of the microscopic model. In particular it has to account for the phase transition which occurs at $T_c(c)$. Accordingly, $V_L(c(\vec{x}))$ is a polynomial in $c(\vec{x})$ of the form :

$$V_L(c(\vec{x})) = -\tfrac{1}{2} \tau_L c^2(\vec{x}) + \tfrac{1}{4!} g_L c^4(\vec{x}) + \dots \quad . \tag{2.22}$$

Above T_c, V_L has only one well, while below T_c, it should have two wells, in order to describe the possibility of two phases. Accordingly, τ_L should be of the form :

$$\tau_L = A_L \frac{(T_c - T)}{T_c} \tag{2.23}$$

with A_L and $g_L > 0$. $\tag{2.24}$

A typical example of V_L is depicted on Fig. 2.8 in Section 2.3. Note that the coarse grained free energy $f_L(c)$ is a well defined function for all values of $c(\vec{x})$. Moreover, below T_c, $V_L(c)$ is not a convex function of c. Accordingly, the chemical potential $\mu_L = \frac{\partial f_L}{\partial c}$ will exhibit a loop similar to the van der Waals loop, as shown in Fig. 1.2. In contrast to the van der Waals loop of mean field theory, however, the shape of this loop depends on the coarse graining size L. Thus $f_L(c)$ contains some information about the metastable and unstable states of the system.

An important question in this coarse grained procedure is the choice of the size of the cell L used in the averaging process (or the choice of the cut-off Λ in Fourier space). Obviously, L must be much larger than the lattice constant a_o (i.e. must be larger than the diameter of a "particle" constituting the alloy) in order for the continuum description to make sense. On the other hand, L should not be much larger than the correlation length ξ of the problem because then the cell could contain two phases and one would have lost the details of the phase separation which one is interested in describing. Finally, in order to include all the critical fluctuations in the step one of the partial trace, it makes sense to choose L = $\gamma\xi$, where γ is a number of order unity.

In the limiting case where L is comparable to the size of the system, then the coarse grained free energy tends towards the exact equilibrium convex free energy discussed in Section 2.1 (see Langer [G1]).

2.3   Calculations of the Helmholtz Free Energy Functional

2.3.1. Monte Carlo Studies

In principle one would like to coarse grain or renormalize the dynamical equations of motion which we discuss in the next chapter. The goal would be to have a partially renormalized equation of motion, in which the coarse graining size is of the order of magnitude of the correlation length ξ. However, as we mention later very little progress has been made so far on this problem. Thus in this section we limit our attention to the calculation of the free energy functional defined in Section 2.2, since this quantity plays an important role in the dynamics.

Rather microscopic coarse graining calculation has been carried out in a Monte Carlo (computer simulation) study of the three dimensional Ising model Hamiltonian for the binary alloy, Eq. (2.8). One divides a simple cubic lattice of $24 \times 24 \times 24$ sites into cells of edge size L, as in (2.17). One then introduces the probability distribution function $P_L(\{c_\alpha\})$ which is the probability for finding cell

1 with an average concentration c_1, cell 2 with an average concentration c_2, etc, with the average concentration in cell α defined by
(2.18). It is usually assumed that such a distribution function can be
given by a lattice version of the Ginzburg-Landau Hamiltonian whose continuum analogue is (2.21), i.e.

$$P_L(\{c_\alpha\}) = \frac{1}{Z} \exp \{\sum_\alpha (h_L c_\alpha - \frac{1}{2} \tau_L c_\alpha^2 + \frac{1}{4!} g_L c_\alpha^4 + \dots)$$

$$+ \sum_{<\alpha\beta>} C_L (c_\alpha - c_\beta)^2 + \dots = \frac{1}{Z} e^{-\beta F_L \{c_\alpha\}} . \qquad (2.25)$$

What one would in principle like is to compute $P_L(\{c_\alpha\})$ for different
choices of cell size L and in particular for $L = \gamma\xi$, with $\gamma = 0(1)$.
This is, however, very difficult to do. A less ambitious, but still useful, project is to do such a calculation for the reduced one and two-
point distribution functions

$$P_{1,L}(c_\alpha) = \int \prod_{\beta \neq \alpha} P_L(\{c_\beta\}) dc_\beta \qquad (2.26)$$

and

$$P_{2,L}(c_\alpha, c_\beta) = \int \prod_{\gamma \neq \alpha, \beta} P_L(\{c_\gamma\}) dc_\gamma . \qquad (2.27)$$

(Note that for finite L the integral signs in (2.26) and (2.27) should
be replaced by sums, since c_β is a discrete variable. However, for L
sufficiently large c_β becomes quasi-continuous, as can be seen from
(2.18), so that the sums can be well approximated by integrals.) An
evaluation of $P_{1,L}(c_\alpha)$ was originally carried out by Binder [9] by Monte
Carlo sampling procedure. More recently the two-point distribution function $P_{2,L}(c_\alpha, c_\beta)$ has been evaluated for the three dimensional Ising
model (2.16) with $h_L = 0$ (for α and β nearest neighbour cells) by
the same method by Kaski, Binder and Gunton [10]. This two-point distribution is of considerable interest because it has the same qualitative features as the full distribution function $P_L(\{c_\alpha\})$. In particular, it can be parametrized in a form similar to (2.25), i.e.

$$P_{2,L}(c_\alpha, c_\beta) = \frac{1}{Z_0} \exp\{\hat{C}_L (c_\alpha - c_\beta)^2 + \hat{V}_L(c_\alpha) + \hat{V}_L(c_\beta)\} \qquad (2.28)$$

$$V_L(c) = -\frac{1}{2} \hat{\tau}_L c^2(\vec{x}) + \frac{1}{4!} \hat{g}_L c^4(\vec{x}) \qquad (2.29)$$

where the coefficients $\hat{C}_L, \hat{\tau}_L, \ldots$ should be reasonably similar to the coefficients $C_L, \tau_L, \ldots$ in (2.25). (The constant Z_0 is an appropriate normalization factor.)

A typical result of the Monte Carlo study of $P_{2,L}(c_\alpha, c_\beta)$ is shown in Fig. 2.6 for the case $L = 6$, $kT = 4.4J$ ($J \equiv J_I$) where the bulk T_c is given by $kT_c \simeq 4.51J$. The two peaked structure shown there reflects the expected features of two phase coexistence. As the cell size increases (from $L = 2$ to $L = 8$) this peak structure becomes sharper and the values of the maxima move closer to the equilibrium values of the magnetization. Although there is no unique way to represent the data, (2.28) and (2.29) should be a reasonable first approximation. The results for the coupling constants are shown in Fig. 2.7 while the double well potential $\hat{V}_L(c)$ is shown in Fig. 2.8. As is to be expected $\hat{\tau}_L(T)$ vanishes at a temperature $T_c(L)$ which is greater than the bulk T_c, due to finite size effects. As L increases, $T_c(L) \rightarrow T_c$, as it should. The double well potential $\hat{V}_L(c)$ becomes more convex as L increases, as one might expect.

The authors also computed a coarse-grained "spinodal curve", $c_s(L,T)$, using the definition $\partial_c^2 \log P_{2,L}(c,c)\big|_{c_s} = 0$. (This is analogous to the definition one would use for the Helmholtz free energy functional $F_L(c)$ given in (2.25), namely $\partial^2 F_L/\partial c^2 = 0$ where $c = \frac{1}{n}\sum_\alpha c_\alpha$, where n is the number of spins.) The results for this coarse-grained spinodal curve are shown in Fig. 2.9, in a natural scaling form. The behavior of this curve for the asymptotic limits $L/\xi \ll 1$ and $L/\xi \gg 1$ can be derived by heuristic arguments which we do not present here. The physically useful coarse-graining size is $L \sim \xi$. It should be noted that in contrast to a mean field theory in which there is a unique thermodynamic spinodal curve, the coarse-grained "spinodal curve" depends on L. We would expect that a spinodal curve as determined from F_L would exhibit the same qualitative features as shown in Fig. 2.9.

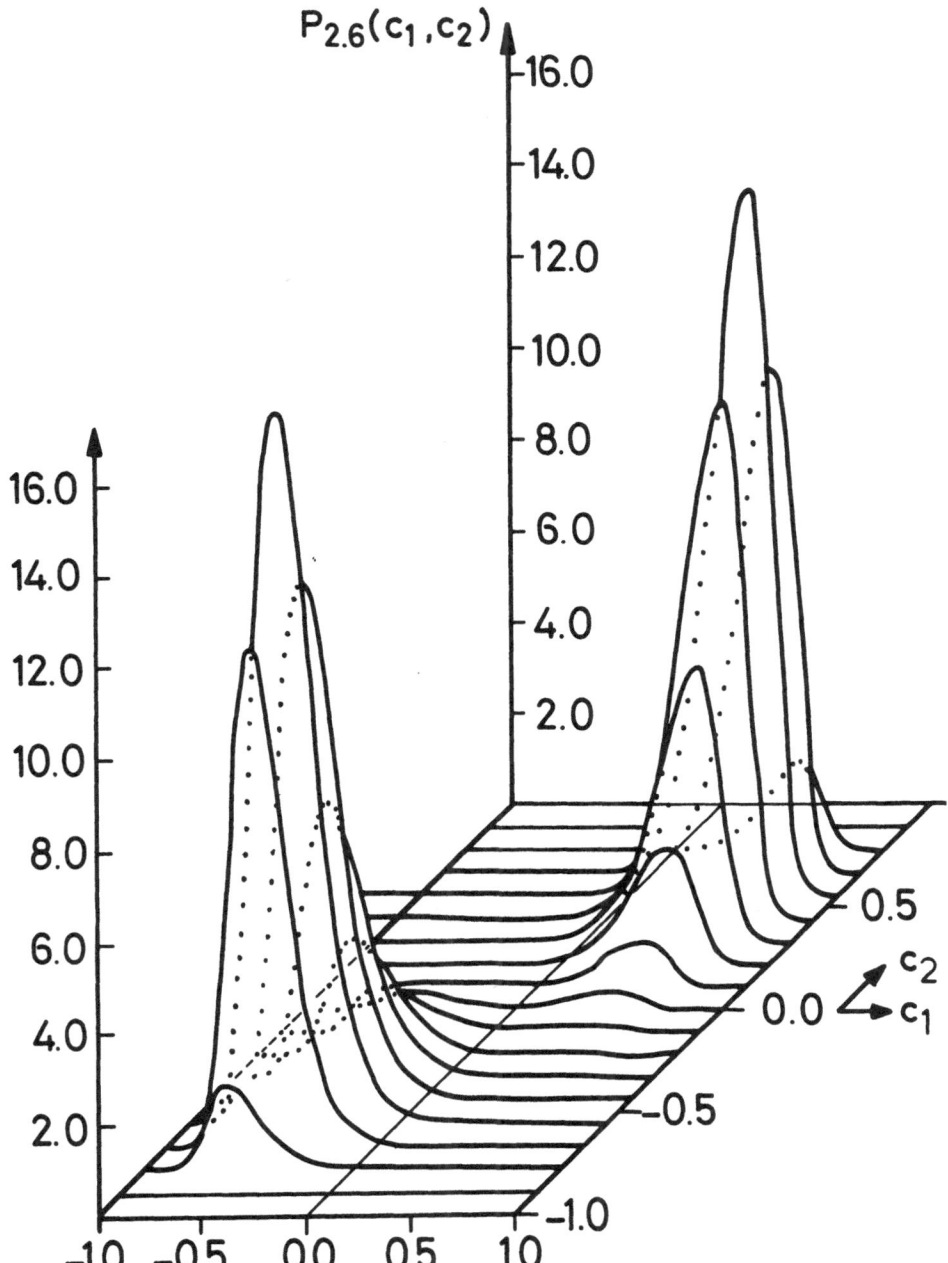

Figure 2.6 : Monte Carlo study of the two-point distribution function
$P_{2,L}(c_1,c_2)$ for the case $L = 6$, $k_BT = 4.4J$, where the
bulk critical temperature T_c is given by $k_BT_c \simeq 4.51J$.

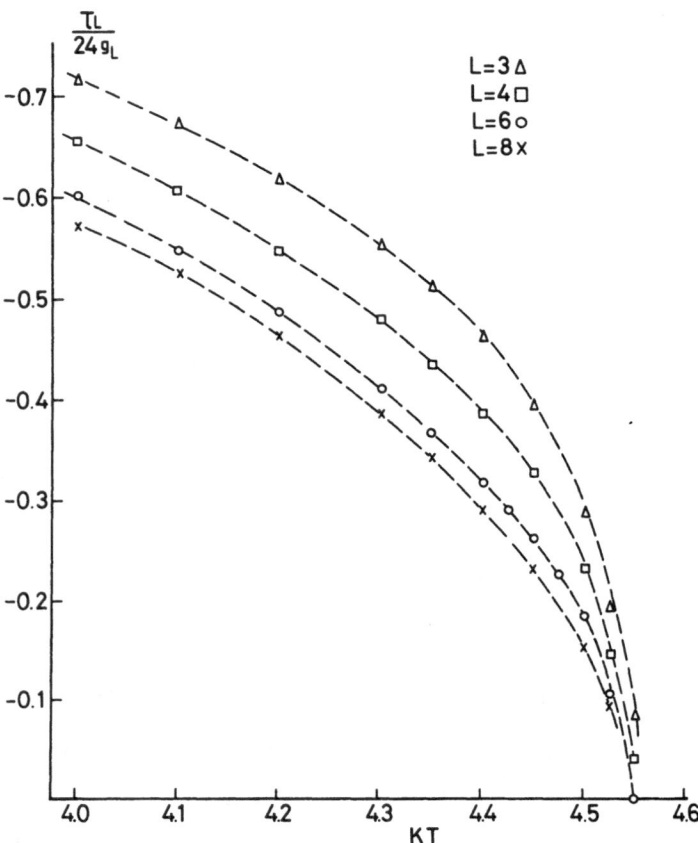

Figure 2.7 : Monte Carlo study of the ratio $\hat{\tau}_L/24\hat{g}_L$ as a function of the temperature for different cell size L.

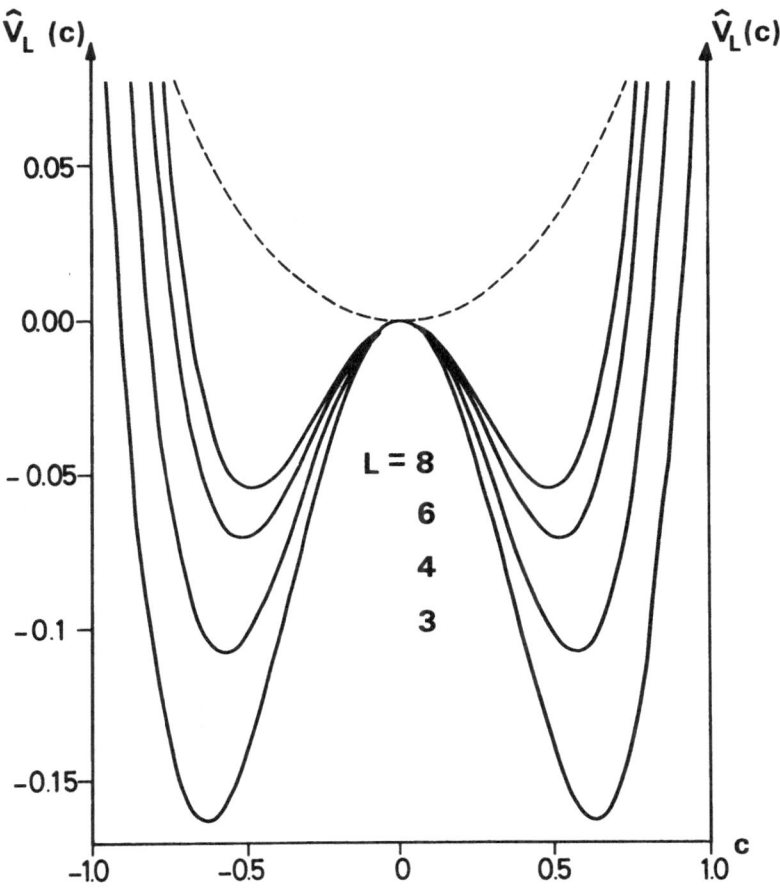

<u>Figure 2.8</u> : Monte Carlo study of the potential $\hat{V}_L(c)$. The full lines
show the double well potentials obtained for $k_BT = 4.45J$
$< k_BT_c$ and different cell size L. The dashed line shows
a typical one well potential obtained for $T > T_c$.

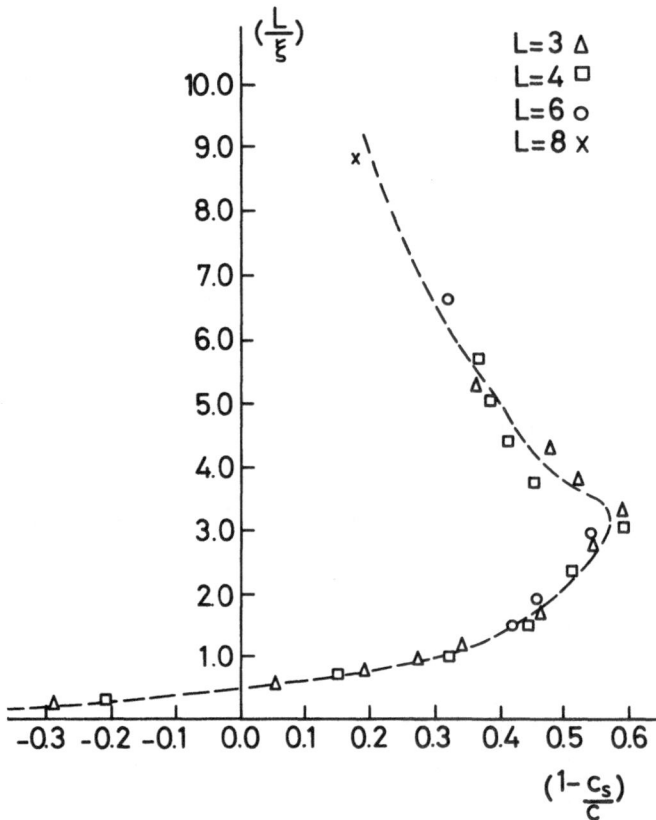

Figure 2.9 : Scaling form for the coarsed-grained spinodal curve.

2.3.2. Field Theory Renormalization Group

It should also be noted that a field theoretic renormalization group calculation of the Helmholtz free energy functional has been carried out to first order in $\epsilon = 4 - d$, by Kawasaki, Imaeda and Gunton [11]. In this approach one starts with the continuum model (2.21), with Fourier components $c_{\vec{k}}$ of $c(\vec{x})$ such that $0 < |\vec{k}| < \Lambda_o$, where Λ_o is some initial cut-off (inverse coarse-graining size). One then averages out fluctuations with Fourier components $k > \Lambda_o e^{-\ell}$, to obtain a new coarse-grained free energy functional with a cut-off $\Lambda_\ell = \Lambda_o e^{-\ell}$. (The quantity e^ℓ is the length rescaling parameter in this momentum space renormalization group.) This procedure is carried out using a differential renormalization group equation of Wegner and Houghton as reformulated by Nicoll, Chang and Stanley [12]. As the de-tails of this procedure are rather complicated, we do not discuss them here. The main results for the coarse-grained free energy functional $F_\ell(c(\vec{x}))$ (as given by a renormalized version of (2.25)) are similar to that discussed in 2.3.1. for the double well potential $\hat{V}_L(c)$, Fig. 2.8. The advantage of this field theoretic calculation is that one can carry out a coarse-graining of the probability distribution function $P_L(\{c(\vec{x})\})$, rather than just the two-point distribution function $P_{2,L}(c_\alpha, c_\beta)$ obtained in the Monte Carlo study. The disadvantage is that the calculation is limited to first order in $\epsilon = 4 - d$, so that the extrapolation to $\epsilon = 1$ necessary to discuss three dimensional systems is rather inaccurate. Further work involving either a real space or a field theoretic calculation of the three dimensional free energy func-tional is clearly necessary. It is of course even more important to carry out such a calculation for the dynamical equations of motion dis-cussed in the next chapter.

REFERENCES - Chapter 2.

[1] C.M. Van Baal, Physica 64, 571 (1973).

[2] K. Huang, "Statistical Mechanics", J. Wiley and Sons Inc.,
     (New York), (1963).

[3] W.L. Bragg and E.J. Williams, Proc. Roy. Soc. 145A, 699 (1934).

[4] K. Binder, Phys. Rev. B8, 3419 (1973).

[5] "Monte Carlo Methods", in Topics in Current Physics, K. Binder ed.,
     Springer-Verlag, Berlin (1979).

[6] C. Domb, in "Phase Transitions and Critical Phenomena", Vol. 3,
     Academic Press (London), edited by C. Domb and M.S. Green (1974).

[7] "Real-Space Renormalization", in Topics in Current Physics, Vol.30,
     Springer Verlag (Berlin), ed. by T.W. Burkhardt and J.M.J. van Leeuwen
     (1982).
     K.G. Wilson and J.B. Kogut, Physics Reports 12C, 77 (1974).

[8] C.J. Thompson, "Mathematical Statistical Mechanics", Princeton
     University Press (Princeton), (1972).

[9] K. Binder, Phys. Rev. Lett. 47, 693 (1981).

[10] K. Kaski, K. Binder and J.D. Gunton, Phys. Rev. Lett. (to be
     published), (1983).

[11] T. Kawasaki, T. Imaeda and J.D. Gunton, in "Perspectives in Sta-
     tistical Physics", M.S. Green Memorial Volume, ed. H.J. Raveche,
     North-Holland (Amsterdam), (1981).

[12] J. Nicoll, T. Chang and H.E. Stanley, Phys. Rev. A13, 1251 (1976).

GENERAL REFERENCES

[G1] J.S. Langer, Physica 73, 61 (1974).

[G2] J.W. Cahn and J.E. Hilliard, J. Chem. Phys. 28, 258 (1958).

[G3] J.W. Cahn and J.E. Hilliard, J. Chem. Phys. 31, 688 (1959).

CHAPTER 3. DYNAMICAL MODEL OF BINARY ALLOYS

In the previous chapter, we defined a model describing the static properties of binary alloys. This model was very simple in the sense that each lattice site was occupied either by an atom of species A or B. This is a very idealized model of a binary alloy. Any real binary alloys will have impurities, vacancies or dislocations. Although these new features may not be relevant for certain static properties of the system, they play an important role in dynamics. Indeed, several mechanisms can contribute to the atomic motion as illustrated in Fig. 3.1.

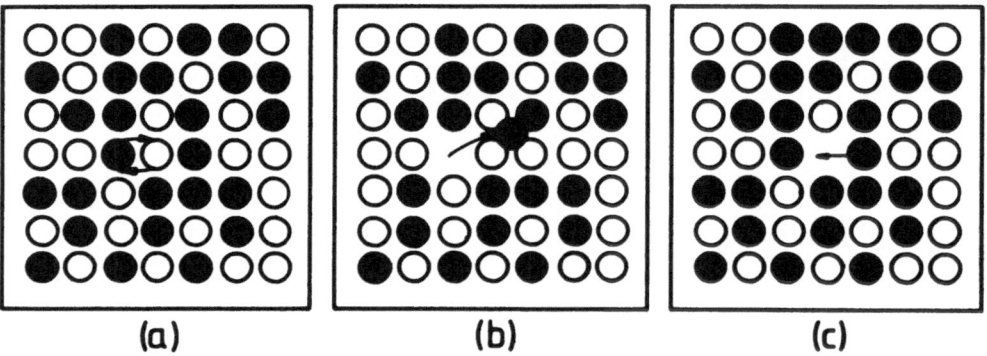

(a)   **(b)**   **(c)**

Figure 3.1 : Three basic mechanisms of diffusion : (a) Interchange by
              rotation about a midpoint. More than two atoms may rotate to-
              gether. (b) Migration through interstitial sites. (c) Atoms ex-
              change position with vacant lattice sites. (From Seitz [6].)

In addition to the simple interchange mechanism of Fig. 3.1a, one can have, for example, migration through interstitial sites (Fig. 3.1b) or interchange with a vacancy (Fig. 3.1c).

In order to be able to describe processes such as nucleation and spinodal decomposition, we have to obtain some model dynamical equations of motion for the relevant quantities describing the system. We aim to derive such equations in the next section.

3.1 Master equation

As we have seen in Section 2.1, the binary alloy can be described microscopically in terms of an Ising-like model for the variables c_i. The dynamics of the variables c_i is given by the usual equations of motion :

$$\partial_t c_i(t) = \{H, c_i(t)\}_- \qquad (3.1)$$

where $\{\ \}_-$ denotes the Poisson bracket (or $-i\hbar$ times the commutator in the quantum case). Using the form (2.8) for the Hamiltonian H, we find immediately that $\partial_t c_i(t) = 0$, i.e. the c_i variables do not have any natural dynamics. However, if one observes experimentally what happens on a given site of the lattice at equilibrium, one sees the following behavior. For some time, the site is occupied by an atom A, then by an atom B and again by an atom A and so on. Obviously the system evolves dynamically.

What is wrong with Eq. (3.1) ? Just the fact that the Hamiltonian H, as given by (2.8), is too simple. As we have seen previously, the model Hamiltonian (2.8) gives a nice description of the static properties of the system. However, it does not take into account all the degrees of freedom.

For example, the degrees of freedom associated with the impurities, vacancies and dislocations discussed previously, as well as the phonons which describe the collective excitations of the lattice, are not taken into account. The interactions resulting between these degrees of freedom and the atoms of the alloy lead to composition fluctuations. It is, however, very complicated (and not necessarily desirable) to give a complete microscopic description of these new interactions. A suitable way to describe their effects consists in simulating them by a heat bath. The composition fluctuations are driven by interactions with the heat bath. The characteristic fluctuation rate is introduced as a phenomenological parameter. Since this dynamical model is phenomenological, we may as well define it for the coarse-grained variables c_α rather than for the c_i.

We are therefore interested in obtaining an equation of motion for the probability density $P(c_1, c_2, \ldots, c_M, t) \equiv P(\{c\}, t)$ that the cell composition configuration $(c_1, c_2, \ldots, c_M)$ is realized at time t. The simplest equation of motion for $P(\{c\}, t)$ that we can write is a master equation, namely :

$$\partial_t P(\{c\}, t) = \sum_{\{c_\alpha\}}' [w(\{c\}, \{c'\}) P(\{c'\}, t)$$
$$- w(\{c'\}, \{c\}) P(\{c\}, t)]$$

(3.2)

where the summation is over all the possible composition configurations $\{c_\alpha\}$. $w(\{c\}, \{c'\})$ is the rate at which the thermal bath induces transitions from $\{c'\}$ to $\{c\}$. The physical source of this heat bath for the binary alloy is presumably the phonon modes, which are assumed to equilibrate rapidly in comparison with the composition changes. The first term on the right hand side of (3.2) expresses the increase of $P(\{c\}, t)$ due to transitions from an arbitrary configuration $\{c'\}$ to $\{c\}$. The second term expresses the decrease of $P(\{c\}, t)$ due to transitions from $\{c\}$ to $\{c'\}$. The major approximation in (3.2) is its Markovian character, i.e., the fact that the change of $P(\{c\}, t)$ is related to $P(\{c'\}, t)$ at the same time only. Thus one assumes that the memory effects are negligible.

Moreover, $P(\{c\}, t)$ satisfies the normalization condition :

$$\sum_{\{c_\alpha\}} P(\{c\}, t) = 1 .$$

(3.3)

The details of the dynamics are contained in the transition rate w. Let us examine more closely what is a reasonable form for w. Since the average composition is conserved, it is reasonable to assume that this average composition is also locally conserved. Thus, if a particular cell composition c_α changes by $+\varepsilon$, we assume that a neighbouring cell α', will have its composition $c_{\alpha'}$ changing from $c_{\alpha'}$ to $c_{\alpha'} - \varepsilon$. The other cells are unaffected by this process. Accordingly, we can write

$$w(\{c'\}; \{c\}) = \frac{1}{2} \sum_{\alpha, \alpha'} \prod_{\beta \neq \alpha, \alpha'} \delta(c'_\beta - c_\beta) D_{\alpha\alpha'} \cdot$$
$$\int_{-\infty}^{\infty} d\varepsilon R(\{c'\}, \{c\}) \delta(c'_\alpha - c_\alpha - \varepsilon) \delta(c'_{\alpha'} - c_{\alpha'} + \varepsilon)$$

(3.4)

where $D_{\alpha\alpha'} = \begin{cases} 1 \text{ if } \alpha \text{ and } \alpha' \text{ are nearest neighbours} \\ \\ 0 \quad \text{otherwise .} \end{cases}$

Note that, strictly speaking, ε can only take some discrete values. However, as the number of sites in a coarse-grained cell is very large, we replace the summation over the values of ε by an integral.

The equilibrium state of the master equation (3.2) satisfies the condition $\partial_t P_e(\{c\}) = 0$. One way to fulfill this condition is to satisfy the detailed balance conditions, i.e.

$$w(\{c\},\{c'\})P_e(\{c'\}) - w(\{c'\},\{c\})P_e(\{c\}) = 0 \qquad (3.5)$$

for all configurations. Moreover, $P_e(\{c\})$ should be proprotional to the Boltzmann factor $\exp - \beta F\{c\}$ (F is the coarse-grained free ener-gy), thus :

$$\frac{w(\{c\},\{c'\})}{w(\{c'\},\{c\})} = \frac{P_e(\{c\})}{P_e(\{c'\})} = \exp - \beta[F\{c\}-F\{c'\}] \ . \qquad (3.6)$$

Accordingly, we write :

$$w(\{c'\},\{c\}) = \exp[\tfrac{1}{2}\beta(F\{c\}-F\{c'\})]\Omega(\{c'\},\{c\}) \qquad (3.7)$$

where $\Omega(\{c'\},\{c\})$ is a symmetric function in the initial and final states, $\{c\}$ and $\{c'\}$ respectively. To determine Ω explicitly, one needs to study the details of the interaction between the heat bath and the variables $\{c_\alpha\}$. We do not do so here. It suffices to remark that, due to the type of transitions allowed by (3.4), Ω is a func-tion only of ε, the change in c. Furthermore, since the coarse-grained cells contain a large number of sites, the change in c cor-responding to a single transition is small. Thus $\Omega(\varepsilon)$ must be sharply peaked around $\varepsilon = 0$ and symmetric in ε. $\Omega(\varepsilon)$ can be characterized by its moments; as we shall see later, only the second moment will be needed. For convenience, we introduce the jump rate per atomic site, Γ, through

$$\int_{-\infty}^{\infty} d\varepsilon \varepsilon^2 \Omega(\varepsilon) = N_c^{-(1+2/d)} \Gamma \ , \qquad (3.8)$$

where N_c is the number of sites in a cell. Thus, Γ is the only adjustable phenomenological parameter of the theory.

Finally, we can rewrite the master Eq. (3.2) in the form :

$$\partial_t P(\{c\},t) = \frac{1}{2} \sum_{\alpha\alpha'}^{'} D_{\alpha\alpha'} \int_{-\infty}^{\infty} d\varepsilon \,\Omega(\varepsilon) \{\exp[\beta\Delta F/2] \cdot$$

$$P(\ldots,c_\alpha+\varepsilon,c_{\alpha'}-\varepsilon,\ldots) - \exp[-\beta\Delta F/2] \cdot$$

$$P(\ldots,c_\alpha,c_{\alpha'},\ldots)\} \tag{3.9}$$

with

$$\Delta F \equiv F\{c_\alpha+\varepsilon,c_{\alpha'}-\varepsilon)\} - F\{c_\alpha,c_{\alpha'}\} \ ,$$

where Σ' means that $c_\beta = c_\beta'$ for $\beta \neq \alpha$ or α'. We can now expand the right hand side in a power series in ε. The different moments of $\Omega(\varepsilon)$ then appear. Taking into account the fact that the odd moments are zero and that the high order moments are very small, we can truncate the expansion and keep only the second moment. The result is

$$\partial_t P(\{c\},t) = \frac{\Gamma}{4N_c^{1+2/d}} \sum_{\alpha\alpha'}^{'} D_{\alpha\alpha'} (\partial_{c_\alpha} - \partial_{c_{\alpha'}}) \cdot$$

$$[\beta(\partial_{c_\alpha} F - \partial_{c_{\alpha'}} F)P(\{c\},t) + (\partial_{c_\alpha} P - \partial_{c_{\alpha'}} P)] \ , \tag{3.10}$$

with $\partial_{c_\alpha} A = \partial A/\partial c_\alpha$.

Moreover, one sees by inspection that (3.10) can be rewritten as

$$\partial_t P(\{c\},t) = \frac{\Gamma}{2N_c^{1+2/d}} \sum_{\alpha} \partial_{c_\alpha} [\beta(2d\partial_{c_\alpha} F$$

$$- \sum_{<\alpha'\alpha>} \partial_{c_{\alpha'}} F)P(\{c\},t) + (2d\partial_{c_\alpha} P(\{c\},t)$$

$$- \sum_{<\alpha'\alpha>} \partial_{c_{\alpha'}} P(\{c\},t))] \ , \tag{3.11}$$

where $\sum_{<\alpha'\alpha>}$ means summation over α' nearest neighbour to α and d is the dimensionality of the system.

Finally, we can write (3.11) in the form of a continuity equation in the {c} space. Defining the probability current vector J_α by :

$$J_\alpha(\{c\},t) = -\sum_\beta \Gamma_{\alpha\beta}[\beta P(\{c\},t)\partial_{c_\beta} F + \partial_{c_\beta} P(\{c\},t)] \qquad (3.12)$$

where

$$\Gamma_{\alpha\beta} = \frac{\Gamma}{2N_c^{1+2/d}} [2d\delta_{\alpha\beta} - \sum_{<\alpha'\alpha>}\delta_{\alpha'\beta}] , \qquad (3.13)$$

the equation (3.11) can be written as :

$$\partial_t P(\{c\},t) = -\sum_\alpha \partial_{c_\alpha} J_\alpha(\{c\},t) . \qquad (3.14)$$

The above derivation is given in terms of the discrete cell composition variables $\{c_\alpha\}$. However, as we have seen in Section 2.2, it is judicious to introduce a continuous field $c(\vec{x})$ which extrapolates smoothly the variables $\{c_\alpha\}$. The continuity equation (3.14) then becomes a Fokker-Planck equation for the time-dependent probability density $P(c(\vec{x}),t)$. In order to derive the Fokker-Planck equation from (3.14), we have to introduce the concept of functional derivative [*] and to find the continuum limit of $\Gamma_{\alpha\beta}$. Formally, the functional derivative is related to the normal derivative by :

$$\partial_{c_\alpha} = (N_c a_o^d) \frac{\delta}{\delta c(\vec{x})} . \qquad (3.15)$$

Moreover the summation on cells becomes an integral over space according to :

$$\sum_\alpha \ldots = (N_c a_o^d)^{-1} \int d^d \vec{x} \ldots . \qquad (3.16)$$

As far as the continuum limit of $\Gamma_{\alpha\beta}$ is concerned, one recognizes in

---

[*] The reader not familiar with functional derivatives can refer to J. Tarski, <u>Functional Integrals in Quantum Field Theory and Related Topics</u>, in "Lectures in Theoretical Physics", Vol. X-A, edited by A.O. Barut and W.E. Brittin, Gordon and Breach, New York (1968).

(3.13) the lattice (or finite difference) version of the second derivative operator [1], i.e.

$$\Gamma_{\alpha\beta} = -\frac{\Gamma}{2} a_o^{d+2} \nabla^2 (\vec{x}-\vec{x}') \ .$$

(3.17)

Using the above identifications in Eq. (3.14), one obtains the desired
Fokker-Planck equation

$$\partial_t P(c,t) = -\int d^d\vec{x} \ \frac{\delta J(c(\vec{x}),t)}{\delta c(\vec{x})}$$

(3.18)

with the probability current

$$J(c(\vec{x}),t) = \frac{\Gamma}{2} a_o^{d+2} \nabla^2 (\vec{x}) [\beta P(c,t) \frac{\delta F}{\delta c(\vec{x})}$$

$$+ \delta P(c,t)/\delta c(\vec{x})] \ .$$

(3.19)

Note that the term ∇^2 in the right hand side of (3.19) expresses the
local conservation of the composition. The Fokker-Planck equation
(3.18) arises in many branches of physics. The above derivation for
this simple model of a binary alloy is due to Langer [2].

3.2   Langevin Equations. Model "A" and "B"

As we have seen with the example of the binary alloy the dynamics
of a system with a very large number of degrees of freedom cannot easily
be described by deterministic equations of motion. However, one can
proceed as follows. One writes down a set of phenomenological equations
of motion for a small number of variables called macrovariables $\{A_i\}$.
These variables are chosen in such a way that they describe correctly
the macroscopic properties of the system. The other degrees of freedom
are supposed to have a much faster evolution and their effect on the
macrovariables is simulated by adding random noise terms in the equations of motion for the A_i's.   Thus :

$$\partial_t A_i(t) = \varphi_i[\{A_j\},t] + \zeta_i(t)$$

(3.20)

where the φ_i's are (generally non linear) functions of all the macrovari-

ables. The functions $\zeta_i(t)$ are random noise terms which are characterized by a probability distribution function. Equations (3.20) are
called Langevin equations. Then one can show that the probability distribution $P(\{A_i\},t)$ satisfies the following Fokker-Planck equation
[3] :

$$\partial_t P[\{A_i\},t] = -\sum_j \frac{\delta}{\delta A_j}[\varphi[\{A_i\}]P] + \sum_j D_j \frac{\delta^2}{\delta A_j^2}P \qquad (3.21)$$

provided that the noise terms have the following properties :

$$<\zeta_i(t)> = 0$$

$$\qquad (3.22)$$

$$<\zeta_i(t)\zeta_j(t')> = 2D_i\delta_{ij}\delta(t-t')$$

and that the $\{D_i\}$ are independent of the macrovariables $\{A_j\}$.

It is sometimes easier to write phenomenological Langevin equations
and then find the associated Fokker-Planck equation, rather than directly deriving the Fokker-Planck equation. Moreover, in order that the
macrovariables describe the correct physics, the various conservation
laws obeyed by the system should be taken into account. Different conservation laws will lead to different Langevin equations. In what follows we shall present two particularly simple models which are called
"model A" and "model B" respectively in the literature (see the review
of Halperin and Hohenberg [4]). Model B is appropriate for the binary
alloy dynamics, Model A, on the other hand, is a model in which the
order parameter is not conserved. It is a model whose dynamical properties will be studied later on in this monograph.

Model A : The macrovariables are the continuum limit $c(\vec{x},t)$ of the
Ising cell variables $c_\alpha(t)$ introduced in Section 3.1. Thus $c(\vec{x},t)$
is the local concentration (or magnetization) field at time t. Model
A considers a situation in which the concentration is not conserved
during the evolution. The Langevin equation for $c(\vec{x},t)$ is :

$$\partial_t c(\vec{x},t) = -\Gamma_o \frac{\delta F}{\delta c(\vec{x},t)} + \zeta(\vec{x},t) \qquad (3.23)$$

with $F = \int d^d\vec{x}[-\frac{1}{2}\tau_o c^2 + \frac{1}{2}c|\nabla c|^2 + \frac{g}{4!}c^4]$ \qquad (3.24) and $<\zeta(\vec{x},t)> = 0$

$$(3.25)$$

$$<\zeta(\vec{x},t)\zeta(\vec{x}',t')> = 2\Gamma_o\delta(\vec{x}-\vec{x}')\delta(t-t') \ .$$

This is a purely relaxational model. The $c(\vec{x},t)$ evolves towards its equilibrium value obtained by extremalizing the free energy F. The associated Fokker-Planck equation is simply :

$$\partial_t P[c(\vec{x},t)] = \Gamma_o \int d^d\vec{x} \ \frac{\delta}{\delta c(\vec{x},t)} [\frac{\delta P}{\delta c(\vec{x},t)} + P\frac{\delta F}{\delta c(\vec{x},t)}] \ . \qquad (3.26)$$

Model B : The Model B differs from Model A in one very important aspect, namely the order parameter, i.e. the concentration $c(\vec{x},t)$, is conserved. The equations defining Model B are identical to (3.23) to (3.25) providing that one makes the substitution

$$\Gamma_o \rightarrow M\nabla^2 \ . \qquad (3.27)$$

Thus

$$\partial_t c(\vec{x},t) = -M\nabla^2(\vec{x}) \ \frac{\delta F}{\delta c(\vec{x},t)} + \zeta(\vec{x},t) \qquad (3.28)$$

with

$$F = \int d^d\vec{x}[-\frac{1}{2}\tau_o c^2 + \frac{1}{2}C|\nabla c|^2 + \frac{g}{4!}c^4] \qquad (3.29)$$

where $<\zeta(\vec{x},t)> = 0$

$$(3.30)$$

and $<\zeta(\vec{x},t)\zeta(\vec{x}',t')> = 2M\nabla^2(\vec{x})\delta(\vec{x}-\vec{x}')\delta(t'-t) \ .$

The associated Fokker-Planck equation is :

$$\partial_t P[c(\vec{x},t)] = M \int d^d\vec{x}\nabla^2(\vec{x}) [\frac{\delta P}{\delta c(\vec{x},t)} + P\frac{\delta F}{\delta c(\vec{x},t)}] \qquad (3.31)$$

This equation is similar to the one obtained in the preceding section for binary alloys. The order parameter relaxation time depends on the wave vector k : $\tau_r(k) \sim k^{-2}$.

3.3 Equations of Motion for the Nonequilibrium Correlation Functions

Knowing the Fokker-Planck equation for the probability distribution $P(c,t)$ we can compute the equal time correlation functions of physical interest. If c is the average composition, and $u(\vec{x},t)$ the composition fluctuation :

$$u(\vec{x},t) = c(\vec{x},t) - c \qquad (3.32)$$

then one is interested in the equal time correlation functions $S_n(\vec{x},t)$ defined as :

$$S_n(\vec{x}-\vec{x}',t) = \langle u^{n-1}(\vec{x},t)u(\vec{x}',t)\rangle$$

$$\equiv \int D[u]u^{n-1}(\vec{x})u(\vec{x}')P(u,t) \qquad (3.33)$$

where $\int D[u]$ denotes the functional integral over the fields $\{u\}$. Note that we have assumed that the above correlation functions are invariant under translations.

Two correlation functions are of particular interest, the one and two-point functions S_1 and S_2.

3.3.1. Equation of motion for the one-point function

S_1 is the average local fluctuation of composition. Its equation of motion is simply obtained by taking the time derivative of Eq. (3.33) with $n = 1$.

$$\partial_t S_1(\vec{x},t) = \int D[u]u(\vec{x})\partial_t P(u,t) . \qquad (3.34)$$

For $\partial_t P$ we use the Fokker-Planck equation (3.18). Thus :

$$\partial_t S_1(\vec{x},t) = -\int D[u]u(\vec{x})\int d^d\vec{y} \frac{\delta J[u(\vec{y}),t]}{\delta u(\vec{y})} .$$

Integrating by parts and taking into account the fact that $J(u(\vec{x}),t)$ vanishes for $u \to \pm\infty$, one finds :

$$\partial_t S_1(\vec{x},t) = \int d^d\vec{y} \, D[u]J(u(\vec{y}),t)\frac{\delta u(\vec{x})}{\delta u(\vec{y})} = \int D[u]J(u(\vec{x}),t) . \qquad (3.35)$$

Substituting $J(u,t)$ by its form (3.19), one gets

$$\partial_t S_1(\vec{x},t) = \frac{\Gamma a_0^{d+2}}{2} \beta \int D[u] \nabla^2(\vec{x})\{P(u,t)\frac{\delta F}{\delta u(\vec{x})}\}$$

$$+ \frac{\Gamma a_0^{d+2}}{2} \int D[u] \nabla^2(\vec{x})\{\frac{\delta P(u,t)}{\delta u(\vec{x})}\} . \qquad (3.36)$$

The second integral vanishes because $P(u,t) \to 0$ for $u \to \pm\infty$. Thus

$$\partial_t S_1(\vec{x},t) = \frac{\Gamma a_0^{d+2}}{2k_B T} \nabla^2(\vec{x}) < \frac{\delta F}{\delta u(\vec{x})} > . \qquad (3.37)$$

Several remarks can be made at this point.

i) Let us suppose that the probability distribution $P(u,t)$ is a sharply peaked function around $u(\vec{x},t)$. Thus in this approximation, any function of $\{u(\vec{x}),t\}$ can be accurately replaced by the same function of $u(\vec{x},t)$. In particular (3.37) can be rewritten as :

$$\partial_t c(\vec{x},t) = \frac{\Gamma a_0^{d+2}}{2k_B T} \nabla^2(\vec{x}) \frac{\delta F}{\delta c(\vec{x},t)} \qquad (3.38)$$

which is just the phenomenological hydrodynamic equation for the conserved quantity c.

ii) Note that if we assume translational invariance for the S_n, (as we did above in (3.33)), then S_1 is strictly zero in which case (3.37) becomes trivial. This assumption of translational invariance is in fact used in the Langer, Barron, Miller theory discussed in Chapter 7.

iii) As we have seen in Section 2.2, the coarse-grained free energy F is a functional of the field $c(\vec{x})$ containing to lowest order a quadratic and a quartic term. Thus, by taking the functional derivative with respect to c, we will generate linear and cubic terms in c. The linear term will give rise to a term proportional to S_1 in (3.37). However, the cubic term will generate a three-point function. Thus, the nonlinearities in F lead to an equation of motion for S_1 which is not closed : three-point functions S_3 appear. (Higher order n-point functions would also appear in this equation for S_1 if one were to include higher order terms in F.) If we were to write the equation of motion for S_3, higher order n-point functions would be generated. Thus, we have to face the problem of solving an infinite hierarchy of coupled equations. As we shall see in the next section, this is a general problem.

3.2.2. Equation of motion for the two-point function

The two-point function $S_2(\vec{x},t)$ plays a particularly important role because its Fourier transform $S_2(\vec{k},t)$ is directly proportional to the scattered intensity $I(\vec{k},t)$ in any scattering experiment [3]. The wave vector $\vec{k}$ is the difference between the wave vectors of the incident and scattered particles.

Let us look at the equation of motion for $S_2(\vec{x},t)$, following the same method used for S_1. The starting equation of motion is

$$\partial_t S_2(\vec{x},t) = \int D[u] u(\vec{x}) u(\vec{0}) \partial_t P(u,t)$$

$$= -\int D[u] u(\vec{x}) u(\vec{0}) \int d^d\vec{y} \frac{\delta J(u(\vec{y}),t)}{\delta u(\vec{y})} \quad . \tag{3.39}$$

Upon integrating by parts and taking into account the boundary conditions, one obtains

$$\partial_t S_2(\vec{x},t) = \int d^d\vec{y} \int D[u] \{\delta(\vec{x}-\vec{y}) u(\vec{0}) + \delta(\vec{y}) u(\vec{x})\} J(u(\vec{y}),t)$$

$$= 2\int D[u] u(\vec{0}) J(u(\vec{x}),t) \quad . \tag{3.40}$$

Replacing J by its expression (3.19) we can write :

$$\partial_t S_2(\vec{x},t) = I_1(\vec{x},t) + I_2(\vec{x},t) \tag{3.41}$$

where

$$I_1(\vec{x},t) = \beta \Gamma a_o^{d+2} \int D[u] u(\vec{0}) \nabla^2(\vec{x}) [P(u,t) \frac{\delta F}{\delta u(\vec{x})}] \quad , \tag{3.42}$$

$$I_2(\vec{x},t) = +\Gamma a_o^{d+2} \int D[u] u(\vec{0}) \nabla^2(\vec{x}) (\frac{\delta P(u,t)}{\delta u(\vec{x})}) \quad . \tag{3.43}$$

The quantity $I_2(\vec{x},t)$ is readily computed by integration by parts. One finds :

$$I_2(\vec{x},t) = -\Gamma a_o^{d+2} \nabla^2(\vec{x}) \delta(\vec{x}) \ . \tag{3.44}$$

To compute I_1, we can expand $F\{u(\vec{x})\}$ in a power series of u. According to (2.21) we have :

$$F\{u(\vec{x})\} = \int d^d \vec{x} \{\tfrac{1}{2} C |\nabla u(\vec{x})|^2 + V(u(\vec{x}))\} \ .$$

But

$$V(u(\vec{x})) = V_o + V_o^{(1)} u(\vec{x}) + \tfrac{1}{2} V_o^{(2)} u^2(\vec{x}) + \ldots$$

$$= \sum_{n=0}^{\infty} \frac{1}{n!} V_o^{(n)} u^n(\vec{x}) \ , \tag{3.45}$$

where $V_o^{(n)} = \delta^n V(u(\vec{x}))/\delta u(\vec{x})^n \big|_{u=0}$.

Thus $\quad \dfrac{\delta F}{\delta u(\vec{x})} = -C \nabla^2(\vec{x}) u(\vec{x}) + \sum_{n=1}^{\infty} \dfrac{n}{n!} V_o^{(n)} u^{(n-1)}(\vec{x})$

and $\quad I_1(\vec{x},t) = \beta \Gamma a_o^{d+2} \nabla^2(\vec{x}) \{-C \bar{S}_2(\vec{x},t) + \sum_{n=2}^{\infty} \dfrac{n}{n!} V_o^{(n)} S_n(\vec{x},t)\} \tag{3.46}$

where $\quad \bar{S}_2(\vec{x},t) = \langle u(\vec{0}) \nabla^2(\vec{x}) u(\vec{x}) \rangle$. $\tag{3.47}$

Thus finally :

$$\partial_t S_2(\vec{x},t) = \beta \Gamma a_o^{d+2} \nabla^2(\vec{x}) \{-C \bar{S}_2(\vec{x},t) + \sum_{n=2}^{\infty} \frac{n}{n!} V_o^{(n)} S_n(\vec{x},t)\}$$

$$- \Gamma a_o^{d+2} \nabla^2(\vec{x}) \delta(\vec{x}) \ , \tag{3.48}$$

or, in terms of the Fourier transform :

$$S_n(\vec{k},t) = \int S_n(\vec{x},t) e^{i\vec{k}\cdot\vec{x}} d^d\vec{x} \ , \tag{3.49}$$

the equation of motion for the two-point function is

$$\partial_t S_2(\vec{k},t) = -\beta \Gamma a_o^{d+2} k^2 \{(V_o^{(2)} + Ck^2 S_2(\vec{k},t)$$

$$+ \sum_{n=3}^{\infty} \frac{n}{n!} V_o^{(n)} S_n(\vec{k},t)\} + \Gamma a_o^{d+2} k^2 \ . \tag{3.50}$$

Here again, higher order correlation functions appear in the right hand side of the equation. Similarly one can derive the equation of motion for S_n with $n > 2$. Besides higher order correlation functions S_m $(m > n)$, new correlation functions of the type

$$S_{a,b}(\vec{x},t) = <u^a(\vec{x})u^b(\vec{0})> \quad \text{appear.}$$

Thus, as already observed in the former section, we have to face the problem of solving an infinite hierarchy of coupled equations. One obviously has to resort to approximate methods to solve these equations. Several of these approximations will be discussed later on.

REFERENCES - Chapter 3.

[1] M. Abramowitz and A. Segun, "Handbook of Mathematical Functions",
Dover (New York), p. 885 (1970).

[2] J.S. Langer, Ann. Phys. 65, 53 (1971).

[3] H. Haken, Rev. Math. Phys. 47, 67 (1975).
Z. Schuss, "Theory and Applications of Stochastic Differential
Equations", John Wiley and Sons, New York (1980).

[4] P.C. Hohenberg and B.I. Halperin, Rev. Mod. Phys. 49, 435 (1977).

[5] A. Guinier and G. Fournet, "Small Angle Scattering of X-Rays",
John Wiley and Sons, New York (1955), p. 28.

[6] F. Seitz, "The Modern Theory of Solids", McGraw Hill, New York
(1940).

As we have seen in the Introduction, if one brings a system into a metastable state, this system does not remain in this state but eventually reaches a true equilibrium state. An example of this is the binary alloy discussed in the previous chapters. If one quenches a given alloy, with concentration c, this alloy will not remain homogeneous. Some nuclei or droplets at one of the equilibrium concentrations will emerge in the sample. The theory of nucleation aims to explain the rate of formation of such nucleating droplets. In what follows, we consider only the case of homogeneous nucleation, i.e. the type of nucleation which takes place in a completely homogeneous phase with no foreign bodies (ions, wall surfaces,...) present. Unfortunately in most solids, nucleation is heterogeneous, and occurs at grain boundaries, dislocations, etc. Thus our discussion of homogeneous nucleation is not applicable to nucleation phenomena in such systems. Nevertheless, it serves as a useful introduction to most of the basic concepts of nucleation theory. An excellent recent review of nucleation phenomena in solids is that of Russell [1].

The purpose of this chapter is to review the main features of the classical theory of nucleation. A good review of this theory is in the book of Abrahams [2]. In Section 4.1 we discuss the equilibrium properties of the classical droplet model. In Section 4.2, we review the Becker-Döring theory [3].

4.1 Equilibrium properties of the classical droplet model

We discuss first the equilibrium properties of the classical droplet model which underlies the dynamical theory. This model provides a useful insight into the mechanism of decay of a metastable state. This model will also serve as a convenient introduction to the field theory model discussed in Chapter 5.

As we have seen in Chapter 2, our binary alloy can be mapped onto an Ising model. Accordingly, we shall use the Ising model language to describe the droplet model. Imagine a lattice with N Ising spins ($S_i = \pm 1$) in a small positive magnetic field h, at a temperature T sufficiently below the critical temperature T_c that almost all the spins are up. If one slowly changes h to negative values, the system will find itself in a metastable state with positive magnetization in a negative field. This situation is depicted in Fig. 4.1 where the magnetization is plotted as a function of the magnetic field. The problem is to explain how such a state eventually decays.

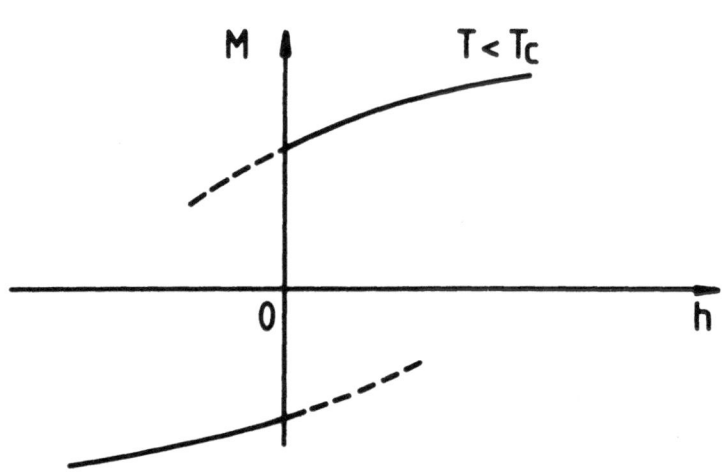

Figure 4.1 : The magnetization M as a function of the magnetic field h for a typical isotherm below T_c. The solid and dashed lines represent the stable and metastable phases respectively.

For positive field h, the typical configurations of this ferromagnet consist of small clusters (or droplets) of down spins dispersed in a background of up spins. Moreover, the average distance between these droplets is sufficiently large such that one can treat the system as a gas of non interacting droplets. The number of droplets of size ℓ is then given by the Boltzmann factor :

$$n_\ell = \aleph e^{-\beta \epsilon_\ell} \qquad\qquad\qquad (4.1)$$

where $\beta = 1/k_B T$ and ϵ_ℓ is the free energy of formation of a droplet of size ℓ. $\aleph$ is a normalization factor. The physics is in ϵ_ℓ. The classical assumption is that ϵ_ℓ is the sum of two terms, a bulk and a surface term. The bulk term corresponds to the energy needed to flip ℓ spins in a field h, i.e., $2h\ell$. The surface term expresses the energy associated with the "surface tension" σ of the droplet. Assuming that the droplets are mostly spherical like, i.e. more compact than ramified, in a d-dimensional space the surface energy is $\sigma \ell^{(d-1)/d}$. Thus

$$\epsilon_\ell = 2h\ell + \sigma \ell^{(d-1)/d} . \qquad\qquad\qquad (4.2)$$

We can then compute n_ℓ as a function of ℓ for the different situations of interest. For $h > 0$, ϵ_ℓ grows linearly in ℓ and accordingly, n_ℓ decreases rapidly with ℓ (see Fig. 4.2).

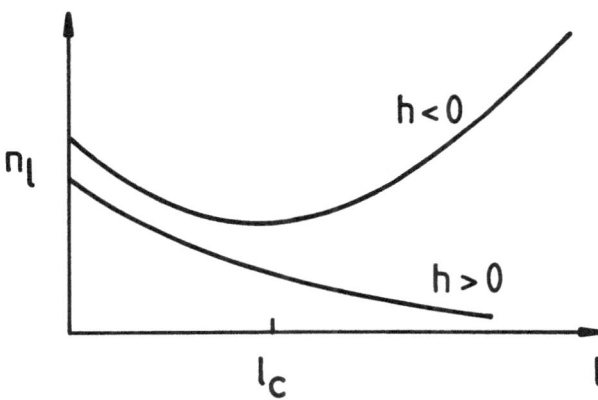

Figure 4.2 : The classical droplet distribution function n_ℓ as a function of the droplet size ℓ, for stable and metastable values of the magnetic field h.

The physical properties are determined by microscopically small droplets. On the other hand, if h is negative, the situation is quite
different. There is a competition between the bulk and the surface term.
The surface term dominates for small ℓ while the bulk term dominates
for large ℓ. As a consequence, there is a critical size droplet ℓ_c
(with a critical radius R_c) such that droplets for which $\ell > \ell_c$ are
energetically favored and grow. These droplets thus provide the nucleating mechanism by which the metastable state decays. From (4.2), we obtain for the critical size ℓ_c :

$$\ell_c = (\frac{\sigma(d-1)}{2d|h|})^d \tag{4.3}$$

4.2 Becker-Döring theory

Since metastability is a dynamical problem, we now consider the
kinetics of cluster formation, as formulated by Becker and Döring
(1935). The starting point of this theory is a kinetic equation for a
time dependent $n_\ell(t)$, where $n_\ell(t)$ is the average number of droplets
of size ℓ present at time t. The basic assumption of their theory
is that the time evolution of $n_\ell(t)$ is only due to an evaporation-
condensation mechanism, in which a droplet of size ℓ loses or gains
a single particle (or an oriented spin in the Ising language). Therefore, effects such as the coagulation of two droplets are not considered, so that the equation of motion for $n_\ell(t)$ can be written as :

$$\partial_t n_\ell(t) = J_{\ell-1}(t) - J_\ell(t), \quad \ell > 2 . \tag{4.4}$$

J_ℓ is the rate per unit volume at which droplets of size ℓ grow to
droplet of size $\ell + 1$. This rate should contain two different mechanisms, the condensation of a particle on a droplet of size ℓ to
form a droplet of size $(\ell + 1)$ and the evaporation of a particle
from the droplet of size $(\ell + 1)$. Assuming that these evaporation
processes are proportional to n_ℓ and $n_{\ell+1}$ respectively, we can
write :

$$J_\ell(t) = R_\ell n_\ell(t) - R'_{\ell+1} n_{\ell+1}(t) , \tag{4.5}$$

where R_ℓ and $R'_{\ell+1}$ are phenomenological kinetic coefficients containing all the details of the kinetic process. Note that Eq. (4.4) does
not hold for single particle clusters, since such clusters are not constrained to events involving other one particle clusters. Following
Becker and Döring, we assume that n_1 remains constant. However, an
alternative treatment has been proposed by Penrose and Lebowitz [4],
who determine n_1 from the conservation of the total number of particles in the system.

The stationary solution of (4.4) is obtained by solving the equation :

$$J_{\ell-1} - J_\ell = 0 . \qquad (4.6)$$

The equilibrium solution is obtained by asking that detailed balance
is fulfilled, i.e., $J_\ell = 0$. Using (4.1) and (4.5), this yields :

$$R_{\ell-1}/R'_\ell = \exp - \beta[\epsilon_\ell - \epsilon_{\ell-1}] . \qquad (4.7)$$

Using (4.5), (4.7) can be rewritten as :

$$J_\ell(t) = R_\ell[n_\ell(t) - \exp\beta[\epsilon_{\ell+1} - \epsilon_\ell]n_{\ell+1}(t)] . \qquad (4.8)$$

It is then suitable to go to a continuous description treating ℓ as
a continuum variable. The equation of motion (4.4) can be written as :

$$\partial_t n_\ell(t) = -\partial J_\ell(t)/\partial\ell \qquad (4.9)$$

with

$$J_\ell(t) = R_\ell[-\frac{\partial n_\ell(t)}{\partial\ell} + n_\ell(t)(1 - \exp(\beta\frac{\partial\epsilon_\ell}{\partial\ell}))] . \qquad (4.10)$$

Thus, the Fokker-Planck equation for n_ℓ reads :

$$\partial_t n_\ell(t) = -\frac{\partial}{\partial\ell}[R_\ell(1 - \exp(\beta\frac{\partial\epsilon_\ell}{\partial\ell}))n_\ell(t) - R_\ell\frac{\partial n_\ell(t)}{\partial\ell}] . \qquad (4.11)$$

Note that this Fokker-Planck equation has an ℓ-dependent diffusion coefficient R_ℓ. Thus, the Becker-Döring equation can be interpreted as a

Markovian stochastic process in ℓ-space [5]. To define completely the theory, we have to specify the kinetic coefficient R_ℓ. The Becker-Döring assumption is that the rate at which molecules condense on a droplet of size ℓ is proportional to its surface area, so that :

$$R_\ell \sim \ell^{(d-1)/d} \; .$$ (4.12)

Let us go back to the stationary condition (4.6). The value $J_\ell = 0$ corresponds to the equilibrium solution. Nucleation theory, on the other hand, is based on a nonequilibrium, steady state solution of (4.11), with $J_\ell = I =$ constant. The quantity I is called the nucleation rate and measures the rate of production of droplets, larger than the critical size, in a nonequilibrium steady state. This non-equilibrium steady state is a time independent solution n_ℓ^s of (4.11). In order to find this solution, one has to specify the boundary conditions for the problem. The boundary conditions usually used are the following :

$$n_\ell^s \to \bar{n}_\ell \; , \quad \ell \to 0 \; ;$$

$$n_\ell^s \to 0 \; , \quad \ell \to \infty \; .$$ (4.13)

This choice of "source and sink" boundary conditions describes the following physical situation. There is a source of droplets at $\ell = 0$ and, once a droplet grows to a specified large size (greater than ℓ_c), it is removed from the system. Moreover, we assume $\bar{n}_\ell$ to be equal to the equilibrium distribution. Thus, the stationary distribution n_ℓ^s is a solution of

$$- \beta R_\ell E(\ell) n_\ell^s + R_\ell \frac{\partial}{\partial \ell} n_\ell^s = -I$$ (4.14)

where

$$\beta E(\ell) = \{ 1 - \exp(\beta \frac{\partial \epsilon_\ell}{\partial \ell}) \}$$ (4.15)

with the boundary conditions (4.13). The general solution of this differential equation involves the following integral :

$$J(\ell) = \int_0^\ell \beta E(\ell') d\ell' \; .$$ (4.16)

By expanding the exponential in (4.15), we have :

$$\beta E(\ell') = -\beta\frac{\partial\epsilon_{\ell'}}{\partial\ell'} - \frac{1}{2}\beta^2(\frac{\partial\epsilon_{\ell'}}{\partial\ell'})^2 + \ldots . \tag{4.17}$$

Taking into account the form (4.2) of ϵ_ℓ, one sees that the first term of expansion dominates $J(\ell)$ and thus, we can write :

$$J(\ell) = -\beta\epsilon_\ell . \tag{4.18}$$

Within this approximation, the general solution of (4.14) reads :

$$n_\ell^s = (\exp-\beta\epsilon_\ell)[-\int_o^\ell \frac{Ie^{\beta\epsilon_{\ell'}}}{R_{\ell'}}d\ell' + C] \tag{4.19}$$

where I and C are two constants. For $\ell \to \infty$, n_ℓ^s should vanish. This is obtained by choosing

$$C = I \int_o^\infty \frac{e^{\beta\epsilon_{\ell'}}}{R_{\ell'}} d\ell' , \tag{4.20}$$

so

$$n_\ell^s = Ie^{-\beta\epsilon_\ell} \int_\ell^\infty \frac{e^{\beta\epsilon_{\ell'}}}{R_{\ell'}} d\ell' . \tag{4.21}$$

For $\ell \to 0$, $n_\ell^s \to \bar{n}_\ell \simeq \exp-\beta\epsilon_\ell$; thus the nucleation rate I is given by :

$$I = \bar{n}_\ell e^{\beta\epsilon_\ell}/\int_o^\infty \frac{e^{\beta\epsilon_m}}{R_m} dm = \frac{1}{\int_o^\infty \frac{dm}{\bar{n}_m R_m}} . \tag{4.22}$$

while the stationary solution reads :

$$n_\ell^s = \frac{\int_\ell^\infty \frac{\bar{n}_\ell}{\bar{n}_m R_m} dm}{\int_o^\infty \frac{dm}{\bar{n}_m R_m}} . \tag{4.23}$$

Note that this steady state solution is at best an approximate description of the real process of nucleation. Several effects present in real processes have been neglected. For example, for a gas-liquid system, the formation of liquid droplets produces a decrease of the initial super-saturation, δc, which implies that the chemical potential difference $\delta\mu = \mu - \mu_c$ is a time dependent quantity. This phenomenon, called the depletion effect, is not taken into account in this steady state solution, since $\delta\mu$ is kept fixed at its initial value.

Let us go back to the Eq. (4.22) for the nucleation rate and make a further approximation. For small initial supersaturation $\exp - \beta\epsilon_\ell$ has a rather sharp maximum at ℓ_c. Thus, the integral (4.22) can be evaluated by expanding the integrand around ℓ_c. One has

$$\epsilon_\ell = \epsilon_{\ell_c} + \frac{1}{2}\left(\frac{\partial^2 \epsilon_\ell}{\partial\ell^2}\right)_{\ell_o} (\ell-\ell_c)^2 + 0((\ell-\ell_c)^3) \; . \tag{4.24}$$

Thus (4.22) yields, with $\epsilon''_{\ell_c} \equiv (\partial^2\epsilon_\ell/\partial\ell^2)_{\ell_c} < 0$,

$$I^{-1} = \frac{1}{R\ell_c N e^{-\beta\epsilon_{\ell_c}}} \int_o^\infty d\ell \; \exp - \frac{1}{2}\beta|\epsilon''_{\ell_c}| (\ell-\ell_c)^2$$

$$= \frac{1}{R\ell_c N e^{-\beta\epsilon_{\ell_c}}} \left(\frac{2}{\beta|\epsilon''_{\ell_c}|}\right)^{1/2} \int_{-x}^\infty dz \; e^{-z^2} \tag{4.25}$$

where $x = [\beta|\epsilon''_{\ell_c}|/2]^{1/2}\ell_c \; . \tag{4.26}$

The integral in (4.25) can be expressed in terms of the error function erf(x). One has :

$$\int_{-x}^\infty dz \; e^{-z^2} = \frac{\sqrt{\pi}}{2}[1 + \text{erf}(x)] \; . \tag{4.27}$$

Thus the expression for the nucleation rate in this approximation is :

$$I = I_o e^{-\epsilon_{\ell_c}/k_B T} \tag{4.28}$$

where

$$I_o = CNR\ell_c \left(\frac{|\epsilon''_{\ell_c}|}{2k_B T}\right)^{1/2} \tag{4.29}$$

and $\quad c^{-1} = \frac{\sqrt{\pi}}{2} [1 + \mathrm{erf}([\beta|\epsilon_{\ell_c}''|/2]^{1/2} \ell_c)]$. $\qquad\qquad\qquad$ (4.30)

This is the Becker-Döring result for the nucleation rate. It describes a thermally activated process. The quantity ϵ_{ℓ_c} is an activation energy (i.e. the energy of formation for the critical droplet). I_o is the so-called nucleation rate "prefactor". It should be noted that the nucleation rate depends very strongly on the exponential term.

REFERENCES - Chapter 4.

[1] K.C. Russel, Advances in Colloid and Interface Science 13, 205
     (1980).

[2] F.F. Abraham, "Homogeneous Nucleation Theory", Academic Press
     (New York and London), (1974).

[3] R. Becker and W. Döring, Ann. der Phys. 24, 719 (1935).

[4] O. Penrose and J.L. Lebowitz, in "Studies in Statistical Mechanics"
     eds. North-Holland (Amsterdam) (1979).

[5] W. Feller, "An Introduction to Probability Theory and Its Applica-
     tions", John Wiley and Sons (New York), Vol. 1 (1968).

GENERAL REFERENCES

[G1] J.W. Cahn, in "Critical Phenomena in Alloys, Magnets and Super-
     conductors", McGraw-Hill, edited by R.E. Mills, E. Ascher and
     R.J. Jaffee (1971). This is a nice discussion of classical nuclea-
     tion theory for fluids.

An enourmous literature dealing with attemps to improve the classical droplet model of nucleation exists. Two references of interest
in this regard are :

[G2] K. Binder and D. Stauffer, Adv. in Phys. 25, 343 (1976). This is
     perhaps the most comprehensive recent summary of a microscopic
     cluster theory of nucleation. More recent work in this area is
     summarized in the review article by Gunton, San Miguel and Sahni,
     Reference [7] in Chapter 1.

[G3] A. Bruce and D. Wallace, Phys. Rev. Lett. 4, (1982). This is an
     interesting renormalization group treatment of the droplet model
     near one dimension. It is particularly interesting because it is
     one of the few (if not the only one) "exact" results in the drop-
     let model literature.

CHAPTER 5. FIELD THEORY OF NUCLEATION : STATICS

As we have seen in the previous chapter, nucleation is characterized by the formation of droplets of critical size. Accordingly, the
static properties of the system should be closely related to the static
properties of the droplets. In Paragraph 5.1, we shall show that, at
low temperature, the static properties of the droplets can be described
by the so-called drumhead model hamiltonian. The content of this model
has a very simple geometrical interpretation. Namely, the free energy
of a droplet is the sum of a surface term, proportional to the surface
tension of the droplet, and a bulk term. In Paragraph 5.1 we shall show
that the drumhead model hamiltonian can be derived from the Ginzburg-
Landau coarse-grained free energy model (2.21). The results obtained
for the drumhead model will be used in Chapter 8 to analyse the dynamics
of interfaces.

In Paragraph 5.2 we shall address ourselves to the problem of
defining a free energy in the metastable phase. We shall show that a
metastable state can be characterized by a complex free energy and
that the imaginary part of this free energy has an essential singularity in the external field h, when h → 0.

5.1 Derivation of the Drumhead Model Hamiltonian

The static properties of our field theory are completely determined by the coarse-grained free energy defined in Chapter 2. Choosing
appropriate length units and omitting the lattice size index L for
the sake of simplicity, one has :

$$F\{c(\vec{x})\} = \int d^d\vec{x} [\frac{1}{2}|\nabla c|^2 - \frac{1}{2}\tau c^2 + \frac{1}{4!}gc^4 - hc] . \qquad (5.1)$$

It is useful in view of future applications to work in a d-dimensional
space. As before, we have chosen the convention $\tau > 0$ for $T < T_c$.
Note that this free energy is invariant under the Euclidean group, i.e.
invariant under the combinations of translations and rotations in the
d-dimensional space.

What we would like ultimately to describe are the properties of droplets. If we have a droplet, we have in the first place an interface. Thus, it is legitimate to see what happens with the simplest possible interface, namely a planar interface. Let us suppose that we impose the following boundary conditions on the system :

$$\lim_{z=\pm\infty} <c(\vec{r},z)> = \pm A \tag{5.2}$$

where $\vec{r}$ denotes the other $(d - 1)$ coordinates. The classical solution is obtained by finding the extremum of the integrand in (5.1), i.e.

$$\frac{\delta F}{\delta c(\vec{x})} = -\frac{\partial^2 c}{\partial z^2} + \frac{\partial}{\partial c}[-\frac{1}{2}\tau c^2 + \frac{1}{4!}gc^4 - hc] = 0 . \tag{5.3}$$

The solution of (5.3) with the above boundary conditions is the well known kink solution, which is, for zero field h,

$$c_{c\ell}(\vec{r},z) \equiv m_{z_0}(z) = m_0 \text{ th } [(\frac{\tau}{2})^{1/2}(z-z_0)] \tag{5.4}$$

with $m_0 = (6\tau/g)^{1/2} .$ (5.5)

Thus (5.3) and (5.5) are compatible if $\tau = \frac{A^2}{6}g$. Moreover, (5.4) defines an interface with a width $\xi \sim \tau^{-1/2}$ centered at $z = z_0$. A sharp interface is obtained in the limit $\tau \to \infty$, with $g \to \infty$ such that m_0 is finite. But τ large and positive means that we are far below the critical temperature T_c, i.e. at low temperature. Note that z_0 is arbitrary. The presence of this interface clearly breaks Euclidean invariance. A breaking of symmetry is generally associated with the presence of Goldstone modes. This is also the case here. To identify those modes, let us go beyond the classical approximation (5.4), by introducing fluctuations. Thus :

$$c(\vec{r},z) = m_{z_0}(z) + \hat{c}(\vec{r},z) . \tag{5.6}$$

The free energy (5.1) becomes :

$$F = F_m + \frac{1}{2} \int d^dx \ \hat{c}(\vec{x})Q\hat{c}(\vec{x}) + 0(\hat{c}^3) \tag{5.7}$$

where F_m is the classical free energy.

The quadratic form can then be diagonalized. As we shall see later, Q has one eigenfunction $\varphi_o(z) = \partial_{z_o} m_{z_o}(z)$ with zero eigenvalue. Thus $\varphi_o(z)$ is the Goldstone mode associated with the breaking of Euclidean symmetry. Indeed, for a small amplitude a, one has :

$$c(\vec{x}) = m_{z_o}(z) + a\varphi_o(z) = m_{z_o}(z) + a\partial_{z_o} m_{z_o}(z) \simeq m_{z_o+a}(z)$$

which represents an interface translated by the amount a. Beside this Goldstone mode, Q has a set of eigenstates with low energies. Namely, the states $\varphi_{\vec{q}}(z) = \exp i\vec{q}\cdot\vec{r} \; \varphi_o(z)$ are eigenstates with eigenvalues q^2. Thus, at low temperature, the modes with small q can be excited and thus, we have to consider field configurations corresponding to a superposition of those modes, i.e.

$$c(\vec{r},z) = m_{z_o}(z) + \sum_{\vec{q}} A_{\vec{q}} \exp i\vec{q}\cdot\vec{r} \; \varphi_o(z) \simeq m_{z_o}(z-f(\vec{r})) \qquad (5.8)$$

where $f(\vec{r}) = \sum_{\vec{q}} A_{\vec{q}} \exp i\vec{q}\cdot\vec{r}$. $\qquad (5.9)$

These configurations represent an interface translated locally by $f(\vec{r})$. It is then natural to work in a local kink fixed coordinate system and to look at configurations :

$$c(\vec{r},z) = m_{z_o}(z-f(\vec{r})) + \eta(\vec{r};z-f(\vec{r})) \qquad (5.10)$$

where η describes the fluctuations around the non-planar interface.

In order to compute the partition function we have to evaluate the functional integral over all the field configurations. This amounts to integrating over all the fluctuations $\hat{c}$ describing the deviations from planar. However, all the fluctuations $\hat{c}$ should not be treated alike. Indeed, the fluctuations leading to a local deformation $f(\vec{r})$ of the interface have already been taken into account by going into the kink-fixed coordinate system. Thus in order to avoid double counting, we have to integrate only over the fluctuations orthogonal to the ones used to construct $f(\vec{r})$. Accordingly, we have to impose the following constraint :

$$\int_{-\infty}^{+\infty} dz \, c(\vec{r},z) \varphi_0 (z-f(\vec{r})) = 0 \ . \tag{5.11}$$

This constraint determines how the kink position $f(\vec{r})$ is specified as a function of the field c. It is assumed that the above equation has an unique solution. Thus configurations with several kinks are not considered. A similar procedure has been proposed by Fadeev and Popov [1] in the context of quantized gauge field theory.

Let us now compute the free energy associated with this interface. We follow here the variational derivation of Kawasaki and Ohta [2], although the result was obtained first by Diehl, Kroll and Wagner [3] in a perturbative framework. In the kink-fixed coordinate system the field can be written :

$$c(\vec{r},z) = \chi(\vec{r},z-f(\vec{r})) \ . \tag{5.12}$$

Expanding for small $f(\vec{r})$, one has

$$c(\vec{r},z) = \chi(\vec{r},z) - f(\vec{r}) \partial_z \chi(\vec{r},z) + 0(f^2) \ . \tag{5.13}$$

Substituting into (5.1), one obtains :

$$F\{c\} = F\{\chi\} + \frac{1}{2}\int d^d\vec{x} [(\partial_{\vec{r}} f)^2 (\partial_z \chi)^2 - 2\partial_{\vec{r}} f \partial_{\vec{r}} \chi \partial_z \chi$$

$$+ 0(f^2, f\partial_{\vec{r}} f)] \ . \tag{5.14}$$

The constraint (5.11) can be rewritten as :

$$\int_{-\infty}^{+\infty} dz \chi(\vec{r},z) \partial_z m_{z_0}(z) = 0 \ . \tag{5.15}$$

In a classical approximation, $\chi(\vec{r},z)$ is simply obtained by extremalizing the integrand in (5.14). Moreover, this extremalization should be performed subject to the constraint (5.15). Using the usual method of Lagrange multipliers, the equation for the extremum is :

$$\lambda\varphi_o(z) = -\frac{\partial F\{c\}}{\partial \chi} = (\partial_z^2\chi + \partial_r^2\chi + \tau\chi - \frac{g}{6}\chi^3)$$

$$+ (\partial_{\vec{r}}f)^2\partial_z^2\chi - 2\partial_{\vec{r}}f\cdot\partial_{\vec{r}}\partial_z\chi - \partial_{\vec{r}}^2f\partial_z\chi . \qquad (5.16)$$

Assuming that $\chi(\vec{r},z)$ varies more rapidly along the z direction than in the perpendicular direction (which is reasonable at low temperature) we can rewrite (5.16) as :

$$\lambda\varphi_o(z) = [(1 + (\partial_{\vec{r}}f)^2)\partial_z^2 + \tau - \frac{g}{6}\chi^2]\chi . \qquad (5.17)$$

Since one expects that $\chi(\vec{r},z)$ is an odd function in z and $\varphi_o(z)$ is an even function in z, λ should vanish. Rescaling the length in the z direction according to :

$$z \rightarrow \bar{z} = z[1 + (\partial_{\vec{r}}f)^2]^{-1/2} \equiv z/a \qquad (5.18)$$

(5.17) becomes :

$$[\partial_{\bar{z}}^2 + \tau + \frac{g}{6}\chi^2]\chi = 0 \qquad (5.19)$$

which is simply the Euler-Lagrange equation associated with (5.1) whose solution is $m(\vec{z})$. Thus :

$$\chi(\vec{r},z) = m(\vec{z}) . \qquad (5.20)$$

Returning to the field c, one has :

$$c(\vec{r},z) = m_{z_o}[(1 + (\partial_{\vec{r}}f)^2)^{-1/2}(z-f(\vec{r}))] \qquad (5.21)$$

and substituting into (5.1) one obtains for the drumhead free energy :

$$F_{dh}(f) = \sigma\int d^{d-1}\vec{r}\{[1 + (\partial_{\vec{r}}f)^2]^{1/2} + 0(f^2,f^4)\} \qquad (5.22)$$

with $\sigma = \int_{-\infty}^{\infty} dz\, \varphi_o^2(z)$. $\qquad (5.23)$

This result has a very simple physical and geometrical interpretation. The energy associated with the interface F_{dh} is simply given by the product of the surface of the interface Σ_d with the surface tension σ. The planar interface is unstable against local deformations but the surface tension tends to restrain these fluctuations. Finally, let us draw the interface (see Fig. 5.1), and consider the points $P = (\vec{r}_o, o)$ and $S = (\vec{r}_o, z)$.

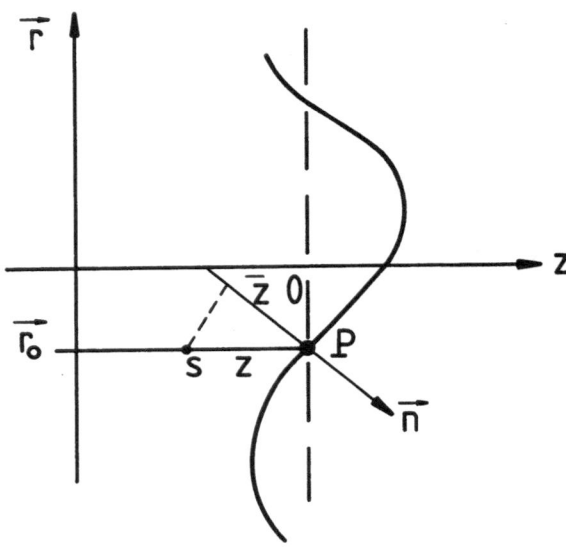

<u>Figure 5.1</u> : A typical interface showing the locally planar nature of the problem. $\vec{n}$ is the normal to the interface at the point P.

Let $\vec{n}$ be the normal to the interface at the point P. The projection z onto the normal $\vec{n}$ is just equal to $\bar{z}$ as defined by (5.18). Thus Eq. (5.21) tells us that the value of the field c at the point S is simply given by the usual classical solution (5.4) at a point which is the projection onto the normal to the interface. Alternatively, consider the non planar interface. Locally, the interface can be approximated by a plane orthogonal to the normal $\vec{n}$. What happens locally in the twisted system of coordinates is similar to what happens globally for the planar interface.

We also see that the drumhead free energy (5.22) is invariant under the Euclidean group. This expresses the fact that the interface can be translated or rotated without cost of energy.

Let us recall once more that the above derivation is a low temperature approximation. If the temperature is too high, short wave length
modes can be excited. Several difficulties may appear. For example, the
deviations from planar can no longer be described by a single valued
function $\delta z = f(\vec{r})$. Some overhangs can occur. Thus it is not clear
how far from the low temperature domain the drumhead model makes sense.

Let us now consider the case of a spherical like interface. The
first step consists in solving the Euler-Lagrange equation associated
with (5.1) for boundary conditions with spherical symmetry. For an arbitrary field h, we do not know the solution. But we are interested
in small fields, (small supersaturation). We know from the droplet model that the radius of the critical droplet is $0(h^{-1})$; thus it is
large for small field. Accordingly, the interface is locally well approximated by a plane. A good approximation for the classical solution
is then [4] :

$$m_{R_o}(R) \simeq \frac{1}{2}(m_+ - m_-) + \frac{1}{2}(m_+ - m_-)\,\text{th}[(\frac{\tau}{2})^{1/2}(R-R_o)] \qquad (5.24)$$

where $m_\pm$ are the constant solutions of the Euler-Lagrange equation,
i.e. :

$$m_\pm = \pm m_o + h/2\tau + 0(h^2) \qquad (5.25)$$

with $m_o = (6\tau/g)^{1/2} \qquad (5.26)$

and R_o is the radius of the droplet.

The boundary conditions correspond to $m(R=0) = m_-$ and
$m(R\rightarrow\infty) = m_+$. Thus (5.24) represents a droplet of the stable phase in
the metastable background. Using (5.25), (5.26), (5.24) can be rewritten as :

$$m_{R_o}(R) = \frac{h}{2\tau} + m_o\,\text{th}[(\frac{\tau}{2})^{1/2}(R-R_o)] \;. \qquad (5.27)$$

The free energy associated with the formation of a droplet can be easily computed within this classical approximation

$$F_{dr}(R_o) = F(m_{R_o}(R)) - F(m_+) . \qquad (5.28)$$

Substituting $m_{R_o}(R)$ and m_+ into (5.1), one finds that :

$$F_{dr}(R_o) = F_{dr}^o(R_o) + hF_{dr}^1(R_o) + 0(h^2) . \qquad (5.29)$$

The zero field part $F_{dr}^o(R_o)$ can be computed by noticing that for low temperature, the integrand is a sharply peaked function around $R = R_o$. Thus, one finds :

$$F_{dr}^o(R_o) = \sigma \Sigma_d(R_o) \qquad (5.30)$$

where the surface tension σ is given by :

$$\sigma = \int_o^\infty dR |\partial_R m_{R_o}|^2 \qquad (5.31)$$

and $\Sigma_d(R_o) = \dfrac{2\pi^{d/2}}{\Gamma(d/2)} R_o^{d-1}$ is the surface of the droplet . $\qquad$ (5.32)

The part linear in the field $F_{dr}^1(R_o)$ is readily computed in the limit $\tau \to \infty$ and leads to :

$$F_{dr}^1(R_o) = 2m_o V_d \qquad (5.33)$$

where $V_d = R_o^d \dfrac{2\pi^{d/2}}{d\Gamma(d/2)} \qquad (5.34)$

is the volume of the droplet.

Thus, taking into account that $h < 0$ for a stable state, we obtain for the energy of the droplet :

$$F_{dr}(R_o) = -2m_o|h|V_d(R_o) + \sigma\Sigma_d(R_o) . \qquad (5.35)$$

The critical radius R_{oc} is obtained by minimizing $F_{dr}(R_o)$ relatively to R_o, which gives :

$$R_{oc} = \frac{\sigma (d-1)}{2m_o |h|} \cdot$$
(5.36)

Finally, the energy needed to create the critical droplet is simply :

$$F_{dr}(R_{oc}) = \frac{2\pi^{d/2}(d-1)^{d-1}}{d\Gamma(d/2)} \cdot \frac{\sigma^d}{[2m_o|h|]^{(d-1)}} \cdot$$
(5.37)

Having the classical results, we can go beyond and take the fluctuations into account. One could extend the derivation of the drumhead model done above for the planar interface to the droplet case. However, the geometrical interpretation given for the quasi-planar interface in conjunction with the results (5.30) and (5.33) yield the following expression :

$$F_{dr} = -2m_o|h|V_d + \sigma\Sigma_d \cdot$$
(5.38)

The problem is then to compute the volume and the surface of the droplet taking into account deviations from a spherical shape. Let us consider the situation as drawn in Fig. 5.2.

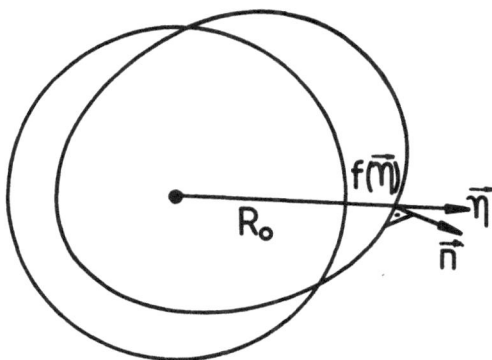

<u>Figure 5.2</u> : Typical deformation of a spherical droplet of radius R_o. $f(\vec{n})$ denotes the displacement in the direction $\vec{\eta}$, while $\vec{n}$ denotes the normal at a point of the deformed surface.

The field $f(\vec{n})$ describes the deviation from a spherical shape in the direction $\vec{\eta}$. $\vec{n}(\vec{\eta})$ is the unit vector normal to the surface of the deformed droplet. The volume is simply given by

$$V_d = \frac{1}{d} \int d\Omega_d (R_o + f)^d \ . \tag{5.39}$$

What about the surface Σ_d ? Let us consider first the two-dimensional case.

$\Sigma_2 = \int_\Gamma ds_2$, where ds_2 is the line element on the closed contour Γ (see Fig. 5.3)

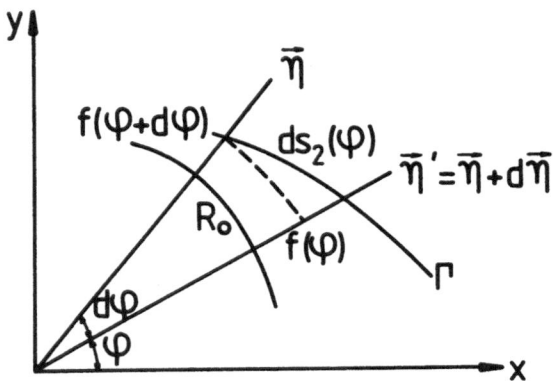

<u>Figure 5.3</u> : An enlarged version of Fig. 5.2 showing the typical geometry involved.

Applying Pythagoras' theorem to $ds_2(\varphi)$, we get :

$$ds_2^2(\varphi) = [(R_o + f(\varphi))^2 d\varphi^2 + (f(\varphi) - f(\varphi + d\varphi))^2]$$

$$= [(R_o + f(\varphi))^2 + (\frac{\partial f(\varphi)}{\partial \varphi})^2] d\varphi^2 \ . \tag{5.40}$$

Thus $ds_2(\varphi) = d\varphi(R_o + f(\varphi))^{(2-1)}[1 + (R_o + f(\varphi))^{-2}(\frac{\partial f}{\partial \varphi})^2]^{1/2} \ . \tag{5.41}$

But $\frac{\partial}{\partial \varphi}$ is just the generator of rotations about the axes z, perpendicular to the plane (x,y). Indeed,

$$\frac{\partial}{\partial \varphi} = x\frac{\partial}{\partial y} - y\frac{\partial}{\partial x} \equiv L_{xy} \tag{5.42}$$

and thus (5.41) becomes :

$$ds_2(\varphi) = d\varphi [R_o + f(\varphi)]^{(2-1)} [1 + (R_o + f(\varphi))^{-2} (L_{xy}f)^2]^{1/2} \qquad (5.43)$$

and

$$\Sigma_2 = \int d\varphi (R_o + f)^{(2-1)} [1 + (R_o + f(\varphi))^{-2} (L_{xy}f)^2]^{1/2} . \qquad (5.44)$$

In d-dimensions, the above construction can be repeated to take into account the deviation from a spherical shape in the other directions. The final result is

$$\Sigma_d = \int d\Omega_d (R_o + f)^{(d-1)} [1 + (R_o + f(\varphi))^{-2} \sum_{i \neq j} (L_{ij}f)^2]^{1/2} . \qquad (5.45)$$

Thus, according to (5.38), (5.39) and (5.45) we can rewritte the droplet energy as :

$$F_{dr}(R_o) = -2m_o |h| \frac{1}{d} \int d\Omega_d (R_o + f)^d$$

$$+ \sigma \int d\Omega_d (R_o + f)^{d-1} [1 + (R_o + f)^{-2} \sum_{i \neq j} (L_{ij}f)^2]^{1/2} . \qquad (5.46)$$

The critical radius R_{oc} is choosen according to the mean field criterion (5.39). Following what we did for the planar drumhead model, we can expand (5.46) up to second order in the fluctuation f. This procedure defines an effective droplet hamiltonian $H_{dh}(R_o)$, given by

$$H_{dh}(R_{oc}) = E_o + \frac{1}{2} \sigma R_{oc}^{d-3} \int d\Omega_d \cdot [\sum_{i \neq j} (L_{ij}f)^2 - (d-1)f^2] \qquad (5.47)$$

where $E_o = \sigma \Sigma_d / d$. $\qquad (5.48)$

Note that (5.47) can be rewritten in a more symmetric form, namely :

$$\int d\Omega_d \sum_{i \neq j} (L_{ij}f \cdot L_{ij}f) = -\int d\Omega_d \sum_{i \neq j} f L_{ij}L_{ij}f + \sum_{i \neq j} f L_{ij}f \Big|_{surface} .$$

$$\qquad (5.49)$$

The second term of the right-hand side is obviously zero and thus :

$$H_{dh}(R_{oc}) - E_o = \frac{1}{2}\sigma R_{oc}^{d-3} \int d\Omega_d f[-\sum_{i \neq j} L_{ij}L_{ij} - (d-1)]f + 0(f^4) . \quad (5.50)$$

The square of the total momentum operator $\mathbb{L}^2$, appears in H_{eff}. The eigenvalues of $\mathbb{L}^2$ are, in d-dimensions, $-\ell(\ell+d-2)$ [5]. Accordingly, the spectrum of (5.50) is given by :

$$\omega_\ell = \frac{1}{2}\sigma R_{oc}^{d-3}(\ell-1)(\ell+d-1) . \quad (5.51)$$

For $\ell = 0$, $\omega_o < 0$, while for $\ell = 1$, $\omega_1 = 0$. Thus the modes with $\ell = 1$ are the Goldstone modes associated with the spontaneously broken invariance. The modes with $\ell > 1$ are spherical harmonic excitations of the surface of the droplet with angular momentum ℓ. Note that expanding (5.46) further leads to higher order terms in $(L_{ij}f)$. These higher order terms do not modify the lowest eigenvalues of the spectrum. As already stated, the drumhead model keeps track only of the low lying excited states which is sufficient a low temperature, but one may expect that this model breaks down when approaching the critical temperature T_c (providing that T_c is not arbitrarily small). We shall come back in the next sections to the physical consequences of the above eigenvalue spectrum. One can also check that the drumhead Hamiltonian is invariant under a non-linear action of the Euclidean group on the field f [6].

5.2 Essential singularity of the free energy

The purpose of this section is to study the behaviour of the free energy in the vicinity of the transition point $h = 0$. We shall see that the free energy has an essential singularity in h, when $h \to 0$. Moreover, we shall see that one can characterize the metastable states by a complex free energy whose imaginary part will be directly related to the nucleation rate in the next chapter.

For the sake of simplicity we shall restrict ourselves to the three dimensional case. However, a similar calculation can be done in arbitrary dimensions (Günther et al.) and we refer the reader to the reference [6] where the d-dimensional case is discussed.

In terms of the coarse grained free energy (5.1), the partition function of the system is :

$$Z = \int D[c] \exp - F\{c\} \qquad (5.52)$$

and the reduced free energy density is :

$$f = \frac{1}{V} \ln Z \ .$$

It is clearly not possible to compute Z exactly in our case. However, it suffices to work within a mean field like approximation to obtain interesting information.

Let us first consider the case of a positive field $h > 0$; i.e. we are in a stable one phase state near the coexistence curve. Similarly to what happens with simple integrals, the functional integral (5.52) is dominated by the configurations $\{c\}$ which maximize the integrand. As we have seen in the previous section, the coarse-grained free energy density has two extrema for $c_{\pm} = m_{\pm}$. The ratio $\exp - F(m_+)/\exp - F(m_-)$ goes to zero exponentially with the volume of the system. Accordingly, the partition function Z is dominated by the configurations $\{c(\vec{x})\}$ close to c_+. Writing $c(\vec{x}) = m_+ + \hat{c}(\vec{x})$, (5.52) becomes :

$$Z \equiv Z_0 = \exp - F(m_+) \int D[\hat{c}] \exp - \frac{1}{2} \int d^3\vec{x} \ \hat{c}(\vec{x}) \, \mathbf{M}_0 \, c(\vec{x}) + 0(\hat{c}^3) \qquad (5.53)$$

with $\quad \mathbf{M}_0 = \left. \dfrac{\delta^2 F}{\delta c(\vec{x})^2} \right|_{c=m_+} . \qquad (5.54)$

From (5.3) we find :

$$\mathbf{M}_0 = -\nabla^2 + \tau + \frac{1}{2} \, gm_+^2 \qquad (5.55)$$

Let $\varphi_j^o(\vec{x})$ and ω_j^o be the eigenfunctions and eigenvalues of the operator $\mathbf{M}_0$. Then $\hat{c}(\vec{x})$ can be expanded in the basis $\{\varphi_j^o(x)\}$:

$$\hat{c}(\vec{x}) = \sum_j \eta_j^o \varphi_j^o(\vec{x}) \; . \tag{5.56}$$

Neglecting the terms of order $\hat{c}^n$, with $n > 2$, one obtains :

$$Z_o = \exp - F(c_+) \int D[\{\eta_j^o\}] \exp - \frac{1}{2} \sum_j \omega_j^o \eta_j^{o2} \; . \tag{5.57}$$

The gaussian integrals can be trivially performed and one gets :

$$Z_o = \exp - F(c_+) \prod_j (\frac{2\pi}{\omega_j^o})^{1/2} \; . \tag{5.58}$$

One could go beyond mean field by treating the higher order terms in $\hat{c}$ in perturbation theory, but this is not necessary to obtain the essential features of the theory.

Let us now consider what happens if we change the field h from its positive value to a slighty negative one such that the average value of c remains positive. This corresponds to changing the system from its stable state to a metastable one. The configurations in the vicinity of $c = m_-$ will characterize the new equilibrium state. The role of the two maxima $m_\pm$ is interchanged. It is therefore reasonable to identify the free energy of the metastable state with the analytic continuation of the free energy associated with Z_o. However, a new feature is now present. Configurations corresponding to the formation of a droplet, i.e. $c(\vec{x}) = m(\vec{R}) + \hat{c}(\vec{x})$ lead to another maximum of the integrand. It turns out, as we shall demonstrate later, that $\exp - F(m_+)/\exp - F(m)$ remains finite when the volume of the system goes to infinity. Thus we have to take into account the contribution of this new maximum when we compute the partition function. Therefore we have

$$Z = Z_o + Z_1 \; . \tag{5.59}$$

One can try to compute Z_1 along the same line used for Z_o and write :

$$Z_1 = \exp - F(m) \int D[\hat{c}] \exp - \frac{1}{2}\int d^3\vec{x} \; \hat{c}(\vec{x}) \, \mathbf{M} \, \hat{c}(\vec{x}) + 0(\hat{c}^3) \tag{5.60}$$

with $M = \dfrac{\delta^2 F}{\delta c(x)^2}\bigg|_{c=m} = -\nabla^2 - \tau + \dfrac{1}{2}gm^2$.

(5.61)

It turns out that the eigenvalues of M have the following properties :
   i) The lowest eigenvalue is negative.
  ii) The second eigenvalue is strictly zero and is three times degen-
      erate. Thus, there are three Goldstone modes.
 iii) All the other eigenvalues are positive.
      Thus we have a saddle point instead of a real maximum.
The negativity of the first eigenvalue make Z_1 purely imaginary. The
second eigenvalue should be treated separately and leads to a term
proportional to the volume V of the system.

   Thus, we can write :

$$Z_1 = iVZ_2$$

(5.62)

and

$$Z = Z_0[1 + iVZ_2/Z_0] .$$

(5.63)

But the free energy should be an extensive quantity. In order to ensure
that, we can exponentiate the relation (5.63) :

$$Z = Z_0 \exp[iV\frac{Z_2}{Z_0}] = Z_0[1 + iV\frac{Z_2}{Z_0} + ...] .$$

(5.64)

The higher order terms in Z_2/Z_0 represent the contributions to Z
comming from the higher order saddle points containing multiple (non
interacting) droplets. Thus, the free energy density f is :

$$f = \frac{1}{V} \ln Z_0 + i\frac{Z_2}{Z_0} .$$

(5.65)

Hence several questions have to be answered and we shall proceed as
follows :
   a) Show that the one droplet configuration contributes even for in-
      finite volume.
   b) Solve the eigenvalue problem associated with M and relate the
      results to the drumhead model.

c) Estimate the contribution of the Goldstone modes to the partition
   function.

d) Compute the free energy density f.

a.   Contribution_of_the_classical_droplet_configuration

The classical droplet configuration $m(R-R_o)$ given by Eq. (5.25)
corresponds to a droplet of a stable phase in a metastable background
$m(R-R_o) \to m_+$ as $R \to \infty$.

Let us compute the ratio

$$\exp - F(m_+)/\exp - F(m) \equiv \exp - \Delta F(m) . \tag{5.66}$$

One has :

$$\Delta F(m) = F(m) - F(m_+) = 4\pi \int_0^\infty dR R^2 [\frac{1}{2}(\nabla m)^2 - \frac{1}{2}\tau (m^2-m_+^2)$$

$$+ \frac{1}{4!}g(m^4-m_+^4) - h(m-m_+)] , \tag{5.67}$$

with the factor 4π coming from the angular integration.

The integral (5.67) can be performed by noticing that for large
τ (i.e. low temperature), the zero field part of the integrant is a
function peaked around $R = R_o$. One then finds easily that :

$$\Delta F(m) = 4\pi\{4\sqrt{2} \ R_o^2\frac{\tau^{3/2}}{g} - \frac{2\sqrt{2}}{\sqrt{3}}|h|\frac{\tau^{1/2}}{g^{1/2}} \ R_o^3\} . \tag{5.68}$$

The surface tension σ defined by (5.31) is thus, in this approximation, given by

$$\sigma = 4\sqrt{2} \ \tau^{3/2}/g . \tag{5.69}$$

The critical droplet radius R_{oc}, is obtained by extremalizing $\Delta F(m)$
with respect to R_o, which leads to :

$$R_{oc} = \frac{4\tau}{\sqrt{3}|h|g^{1/2}} \tag{5.70}$$

and $\quad \Delta F_c = \Delta F(R_o = R_{oc}) = \dfrac{256\pi}{q} [\dfrac{\tau^{7/2}}{|h|^2 g^2}]$ (5.71)

ΔF_c is the activation energy of the critical droplet. Thus the factor $\exp - \Delta F(m)$ does not go to zero when the volume of the system goes to infinity. Therefore we have to take into account this extremum when computing the partition function.

b. Eigenvalue_problem_for_ M

Let us study the eigenvalue problem associated with the operator $\mathbf{M}$ defined by (5.61). Using the spherical symmetry of the problem, one gets :

$$\mathbf{M} = -\dfrac{d^2}{dR^2} - \dfrac{2}{R} \dfrac{d}{dR} + \dfrac{\ell(\ell+1)}{R^2} - \tau + \tau V(R) - hW(R)$$ (5.72)

where $\quad V(R) = 3\mathrm{th}^2 [(\dfrac{\tau}{2})^{1/2}(R - R_{oc})]$ (5.73)

$$W(R) = -\dfrac{\sqrt{6}}{4} (\dfrac{g}{\tau})^{1/2} \mathrm{th}[(\dfrac{\tau}{2})^{1/2}(R - R_{oc})]$$ (5.74)

$\ell = 0,1,2,\ldots$ are the eigenvalues of the angular momentum operator.

In the limit $\tau \to \infty$, $h \to 0$, the term $hW(R)$ can be neglected in (5.72). We are now solving the eigenvalue equation :

$$\mathbf{M}\varphi_{n,\ell}(R) = \omega_{n,\ell}\varphi_{n,\ell}(R) \; .$$ (5.75)

Note that the potential $\tau V(R) - \tau$ has a deep well for $R \simeq R_{oc}$. Accordingly, the lowest eigenstates will be localized at $R \sim R_{oc} \sim h^{-1}$. Thus, the terms in $1/R$ and $1/R^2$ are of order h and h^2 respectively. Provided that ℓ is not too large, we can neglect these terms in first approximation. Thus the eigenvalue equation becomes :

$$\{-\dfrac{d^2}{dR^2} - \tau + 3\tau \, \mathrm{th}^2 [(\dfrac{\tau}{2})^{1/2}(R - R_o)]\}\varphi_{n,\ell}(R) \simeq \omega_{n,\ell}\varphi_{n,\ell}(R) \; .$$ (5.76)

The lowest eigenvalue $\omega_{o,\ell}$ and the corresponding eigenvector $\varphi_{o,\ell}(R)$ are (up to corrections of order $\ell^2 R_{oc}^2$)

$$\omega_{o,\ell} = 0 \qquad\qquad\qquad (5.77)$$

$$\varphi_{o,\ell}(R) = \frac{\tau^{1/4}}{(4\sqrt{2}\pi R_{oc})^{1/2}} (\tfrac{3}{4})^{1/2} \mathrm{sech}^2[(\tfrac{\tau}{2})^{1/2}(R-R_{oc})] \ . \qquad (5.78)$$

The next eigenvalue $\omega_{1,\ell}$ and the corresponding eigenvector $\varphi_{1,\ell}(R)$ are:

$$\omega_{1,\ell} \simeq \frac{3\tau}{4} \qquad\qquad\qquad (5.79)$$

$$\varphi_{1,\ell} = \frac{\tau^{1/4}}{(4\sqrt{2}\,R_{oc})^{1/2}} (\tfrac{3}{2})^{1/2} \frac{\sinh((\tfrac{\tau}{2})^{1/2}(R-R_{oc}))}{\cosh^2((\tfrac{\tau}{2})^{1/2}(R-R_{oc}))} \ . \qquad (5.80)$$

One can show that there are no other bound states.

The modes $\varphi_{o,\ell}$ have an energy close to zero and thus will dominate the partition function. Accordingly, we have to study them in more details. Let us first show that the modes $\varphi_{o,1}(R)$ have a strictly zero eigenvalue. For this purpose, let us consider the Euler-Lagrange equation which has $m(R-R_o)$ as a solution.

In spherical coordinates, this equation is :

$$-\frac{1}{2}\frac{d^2 m}{dR^2} - \frac{1}{R}\frac{dm}{dR} - \frac{\tau}{2}m + \frac{g}{12}m^3 - h = 0 \ . \qquad (5.81)$$

Differentiating with respect to R, we obtain :

$$\{-\frac{1}{2}\frac{d^2}{dR^2} - \frac{1}{R}\frac{d}{dR} + \frac{1}{R^2} - \frac{\tau}{2} + \frac{g}{4}m^2\}\frac{dm}{dR} = 0 \ . \qquad (5.82)$$

A comparison of (5.94) with (5.84), (5.87) leads to the conclusion that

$$\varphi_{o,1}(R) \sim \frac{dm}{dR} \qquad\qquad\qquad (5.83)$$

and $\omega_{o,1} = 0$. $\qquad\qquad\qquad (5.84)$

We have thus proven the assertion made in Section 5.1 about the existence of Goldstone modes restoring the Euclidean invariance of the system. Indeed the fact the $\varphi_{o,1}$ have a zero eigenvalue simply expresses the fact that the partition function is left invariant by a translation of the droplet. Under such a translation

$$m(\vec{R}-\vec{R}_o) \rightarrow m(\vec{R}-\vec{R}_o-d\vec{R}_o) = m(\vec{R}-\vec{R}_o) - d\vec{R}_o \cdot \text{grad } m(\vec{R}-\vec{R}_o) \tag{5.85}$$

and

$$d\vec{R}_o \cdot \text{grad } m(\vec{R}-\vec{R}_o) = \sum_{m=-1}^{+1} \varphi_{o,1}(R) Y_{1,m}(\theta,\varphi) d\eta_{o,1,m} \tag{5.86}$$

where $Y_{\ell,m}(\theta,\varphi)$ are the usual spherical harmonics.

For small ℓ, and $h \rightarrow 0$, the comparison of Eq. (5.82) with (5.72), (5.75) leads to :

$$\omega_{o,\ell} \simeq \frac{\ell(\ell+1)-2}{2R_{oc}^2} = \frac{3h^2 g}{32\tau^2}[\ell(\ell+1)-2] \ . \tag{5.87}$$

Note then that :

$$\omega_{o,o} = -\frac{3h^2 g}{16\tau^2} < 0 \tag{5.88}$$

i.e. the ground state energy is negative. This is the only negative eigenvalue. Apart from the translational modes with zero energy discussed previously, all the other modes $\varphi_{o,\ell}$ with $\ell > 2$ have positive eigenvalues and thus describe stable distortions of the critical droplet.

We can now compare the above results with those of the drumhead model (see Eq. (5.51)). We see that the energy spectrum $\{\omega_\ell\}$ of the drumhead model corresponds to the eigenvalues $\{\omega_{o,\ell}\}$. Thus the drumhead model does not describe all the other modes $\varphi_{n,\ell}$ with $n > 1$. However, as we shall see later, it suffices to keep the modes $\varphi_{o,\ell}$ to obtain the universal properties of the free energy. However, non universal quantities such as the amplitudes, will generally differ if they are computed by a) taking all the modes into account or b) only the $\varphi_{o,\ell}$ modes into account.

c. <u>Contribution of the Goldstone modes to the partition function</u>

The contribution to Z_1 of the Goldstone modes cannot be computed by simple Gaussian integration for obvious reasons. These modes have to be treated separately. Going back to (5.60), we see that to estimate the contribution of the Goldstone modes to Z_1 requires computing the "volume" spanned by those modes in the space of the functions $\{c(\vec{x})\}$.

To estimate this volume, let us first consider a one dimensional case. The Goldstone mode is there $\varphi_{o,1}(R_o) = \dfrac{dm(R-R_o)}{dR}$. As the parameter R_o varies, $\varphi_{o,1}(R_o)$ traces out a line in the function space $\{c(R)\}$. Assume moreover that the system has a length L and that there are periodic boundary conditions such that :

$$m(R-R_o) = m(R-R_o-L) . \tag{5.89}$$

This means that the line will close upon itself. The "length" of a line element is :

$$|\delta\Gamma| = dR_o[\int_o^L(m(R-R_o-dR_o) - m(R-R_o))^2 dR]^{1/2}$$

$$= dR_o[\int_o^L(\frac{\partial m(R-R_o)}{\partial R})^2 dR]^{1/2} . \tag{5.90}$$

Thus the "length" of the closed line is :

$$\Gamma_1 = 2\oint|d\Gamma| \tag{5.91}$$

where the factor 2 takes into account that the $\pm\varphi_{o,1}$ are Goldstone modes in the function space $\{c(R)\}$.

The extension to the three dimensional case is now easy. Indeed one has three orthogonal line elements $|d\Gamma_i|$, $i = 1,2,3$ such that

$$|d\Gamma_i| = dR_{oi}[\frac{1}{3}\int d^3\vec{x}|\vec{\nabla}m(\vec{x}-\vec{R}_o)]^{1/2} \tag{5.92}$$

and thus the desired volume in the function space is :

$$\Gamma_3 = \int|d\Gamma_1 d\Gamma_2 d\Gamma_3| = [\frac{1}{3}\int d^3\vec{x}(\vec{\nabla}m)^2]^{1/2}\int d^3\vec{R}_o$$

$$= V[\frac{1}{3}\int d^3\vec{x}(\vec{\nabla}m)^2]^{1/2} \tag{5.93}$$

where V is the volume of the system.

This last integral can be performed easily using the same approximations made to compute the activation energy (5.68). One finds :

$$\Gamma_3 = V(\frac{32\pi^{1/2}2^{1/4}}{3})^3 (\frac{\tau^{7/4}}{g|h|})^3 \ . \tag{5.94}$$

d. <u>Computation of the free energy density</u>

We can now regroup the different results obtained. From (5.58), (5.60) and (5.66) we obtain upon performing the Gaussian integrals

$$Z_1 = \exp - F(m_+) \ \exp - \Delta F(m) \ \Gamma_3 (\frac{2\pi}{\omega_{oo}})^{1/2} \underset{n,\ell}{\Pi'} (\frac{2\pi}{\omega_{n,\ell}})^{(2\ell+1)/2} \tag{5.95}$$

where the eigenvalues ω_{oo} and ω_{o1} are excluded from the product Π'. The exponent $(2\ell+1)$ takes care of the degeneracy of the modes $\varphi_{n,\ell}$. Since ω_{oo} is negative, Z_1 is purely an imaginary quantity. Moreover, using (5.58) and (5.88), we obtain :

$$Z_2/Z_o = -i\frac{1}{V}\frac{Z_1}{Z_o} = C\frac{\tau^{25/4}}{g^{7/2}|h|^4} \ \exp \ \Phi \tag{5.96}$$

where $C = -\frac{4}{\sqrt{3}}(\frac{32\pi^{1/2}2^{1/4}}{3})^3$ \tag{5.97}

and $\exp\Phi = \underset{n,\ell}{\Pi'} (\frac{2\pi}{\omega_{n,\ell}})^{(2\ell+1)/2} \underset{j}{\Pi} (\frac{\omega^o_j}{2\pi})^{1/2} \ . \tag{5.98}$

Finally, going back to (5.65) we obtain for the imaginary part of the free energy density :

$$\text{Imf} = C\frac{\tau^{25/4}}{g^{7/5}|h|^4} \ \exp \ \Phi \ . \tag{5.99}$$

It remains to compute $\exp \Phi$. This is a tedious calculation and we shall only give here the main steps and hypotheses used. We refer the reader to the original works of Langer [7] and Günther et al. [6] for the details.

Returning to (5.98), we have :

$$\phi = \sum_{n,\ell}' \left(\frac{2\ell+1}{2}\right) \ln\left(\frac{2\pi}{\omega_{n,\ell}}\right) + \sum_{j} \frac{1}{2} \ln\left(\frac{\omega_j^o}{2\pi}\right) \qquad (5.100)$$

where the eigenvalues ω_{oo} and ω_{o1} are excluded from the $\sum'$. The eigenvalues $\omega_{n,\ell}$ are composed by the set $\omega_{o,\ell}$ defined by (5.87) and by the eigenvalues $\{\omega_{n,\ell}, n \geq 1\}$.

Only the eigenvalues $\{\omega_{o,\ell}\}$ are going to zero for $h \to 0$ and thus dominate the sum $\sum'$ in (5.100). As far as the eigenvalues ω_ℓ^o are concerned, one sees from (5.55) that the eigenvalues and eigenfunctions of M_o are (up to a shift 4τ) the ones corresponding to the solution of the Schrödinger equation of a free particle in spherical coordinates. Thus, with $j = (E,\ell,m)$

$$\omega_j^o = 4\tau + E \qquad (E > 0) \qquad (5.101)$$

and $\varphi_j^o(\vec{x}) = j_\ell(E^{1/2}r) Y_{\ell m}(\theta,\varphi) \qquad (5.102)$

where the j_ℓ are the usual Bessel functions of fractional order.

In order to obtain a finite value for ϕ, the sums on ℓ have to be cut-off. The cut-off value ℓ_m is chosen as the one for which the approximation (5.87) for $\omega_{o\ell}$ breaks down, i.e. $\ell_m = bR_{oc} \sim |h|^{-1}$, where $b = 0(1)$. Note also that the shortest meaningful wave length of a fluctuation is given by the lattice spacing. Thus $E << 4\tau$ and $\omega_j^o \simeq 4\tau$.

Finally, one has to extract the asymptotic behaviour of the sums in (5.100) for a large upper cut-off $\ell_m \sim |h|^{-1}$. This can be done by using the Euler-Maclaurin summation formula which states that :

$$\sum_{\ell=n}^{N} f(x_\ell) = \int_n^N f(x)\,dx + \frac{1}{2}[f(x_n) + f(x_N)] +$$

$$\sum_{m=1}^{\infty} \frac{B_{2m}}{(2m)!} [f^{(2m-1)}(x_N) - f^{(2m-1)}(x_n)] \qquad (5.103)$$

where the B_k are the Bernouilli's numbers.

After some tedious algebra, one finds :

$$\exp\phi = D[\frac{\tau^{5/4}g^{5/6}}{|h|^{5/3}}]\exp\{-\frac{\tau^{1/2}}{g}[B(\frac{\tau^{3/2}}{|h|g^{1/2}})^2 + 0(h)]\} \cdot$$

$$\cdot\{1 + 0(g/\tau^{1/2})\} \qquad (5.104)$$

where B and D are dimensionless constants. Note that the particular choice of the parameter b does not affect the form of the singularity in (5.104) as it arises from modes with small ℓ.

Finally, one has for the limit $|h| \to 0$

$$\text{Imf} \sim A\tau^{3/2}(\frac{\tau^{1/2}}{g})^{3/2}(\frac{\tau^{3/2}}{|h|g^{1/2}})^{7/3} \cdot$$

$$\cdot \exp\{-\frac{\tau^{1/2}}{g}[B(\frac{\tau^{3/2}}{|h|g^{1/2}})^2 + 0(|h|)]\}\{1 + 0(\frac{\tau^{1/2}}{g})^{-1}\}. \qquad (5.105)$$

Hence we see that the imaginary part of the free energy has a very weak essential singularity when $h \to 0$. The analytic structure of the function f(h) in the complex h plane can be briefly summarized as follows. f(h) has a branch cut on the negative real axis. The discontinuity of f(h) across the cut is 2Imf(h). The free energy for h > 0 can be obtained through a dispersion relation of the form :

$$f(h) = \frac{1}{\pi}\int_{h'<0} dh' \frac{\text{Imf}(h')}{h'-h} \cdot \qquad (5.106)$$

This yields an essential singularity of f(h) of the same type as the one for Imf. Moreover, all the derivatives of f(h) with respect to h go to zero as $h \to 0$. Thus it may seem that the results derived are not very relevant since they are non observable. We shall see in the next chapter, however, that the situation is not so bad, since Imf is directly proportional to the (measurable) nucleation rate.

A few interesting remarks can be made about equation (5.106). Looking at the d-dimensional generalization of this result, Günther et al. pointed out that the result (5.106) is much more universal than one might anticipate. All the factors appearing in (5.106) are associated

with simple geometrical features of a droplet of radius $R_{oc} \sim |h|^{-1}$.
They do not depend on the details of the coarse-grained free energy we
started with, providing that we restrict ourselves to sufficiently low
temperature. We also see that the field theory developed above is main-
ly an one length scale theory $R_{oc} \sim |h|^{-1}$.

REFERENCES - Chapter 5

[1] L. Fadeev and V. Popov, Phys. Lett. 25B, 29 (1969).

[2] K. Kawasaki and T. Ohta, Prog. Theor. Phys. 67, 147 (1982).

[3] H.W. Diehl, D.M. Kroll and H. Wagner, Zeit. Phys. B36, 329 (1980).

[4] J.W. Cahn and J.E. Hilliard, J. Chem. Phys. 28, 258 (1958);
     J. Chem. Phys. 31, 688 (1959).

[5] E. Madelung, "Die Mathematischen Hilfsmittel Des Physikers", Band
     4, Springer Verlag (Berlin), (1964).

[6] N.J. Günther, D.A. Nicole and D.J. Wallace, J. Phys. A13, 1755
     (1980).

[7] J.S. Langer, Ann. Phys. (N.Y.) 41, 108 (1967).

[8] J.S. Langer, Ann. Phys. (N.Y.) 54, 258 (1969).

GENERAL REFERENCES

[G1] I. Affleck (1980), Ph.D. thesis, Harward University (unpublished).
     This thesis includes a detailed analysis relevant to References
     6, 7 and 8 above.

The goal of this chapter is to compute the nucleation rate within the field theory framework. The method used is a generalization to field theory of the method used for the classical theory of nucleation. This approach was proposed by Langer [1] and involves some earlier work of Landauer and Swanson [2]. The main idea of the calculation can be understood fairly simply knowing the static properties derived in the preceeding chapter. As we have seen, the various stable and metastable configurations occur near the position of local minima of the coarse-grained free energy F{c}. The transitions we are looking for occur when the system starts at one such local minimum and makes its way to another minimum of lower energy. But, in passing from one minimum to the other, the system is more likely to find itself in a state $c = m(R)$ corresponding to a saddle point for F{c} and describing a droplet. Once the saddle point configuration is reached, the system can go from one configuration to the other by continuously lowering its energy until it reaches its equilibrium state. This picture of the trajectory involved in the decay of a metastable state is a crucial assumption of the Langer theory. It is then natural that the saddle point describes the fluctuations which nucleate the phase transition. Generalizing what we did for the Becker-Döring theory, we are looking for a steady state solution of the Fokker-Planck equation which describes a finite probability current flowing across the saddle point. To have that, we have to choose the boundary conditions such that the metastable state is continuously replenished at a rate equal to the one at which it is leaking across the activation-energy barrier.

Let us return to the Fokker-Planck equation (3.18) :

$$\partial_t P\,(c(\vec{x}),t) \;=\; -\int d^3x \;\frac{\delta J((c(\vec{x}),t)}{\delta c(\vec{x})} \tag{6.1}$$

with (in appropriate units)

$$J\,(c,t) \;=\; \Gamma \nabla^2(\vec{x}) \;\left(\frac{\delta F}{\delta c(\vec{x})}P \;+\; \frac{\delta P}{\delta c(\vec{x})}\right) \tag{6.2}$$

We are looking for a stationary solution of (6.1) and (6.2).

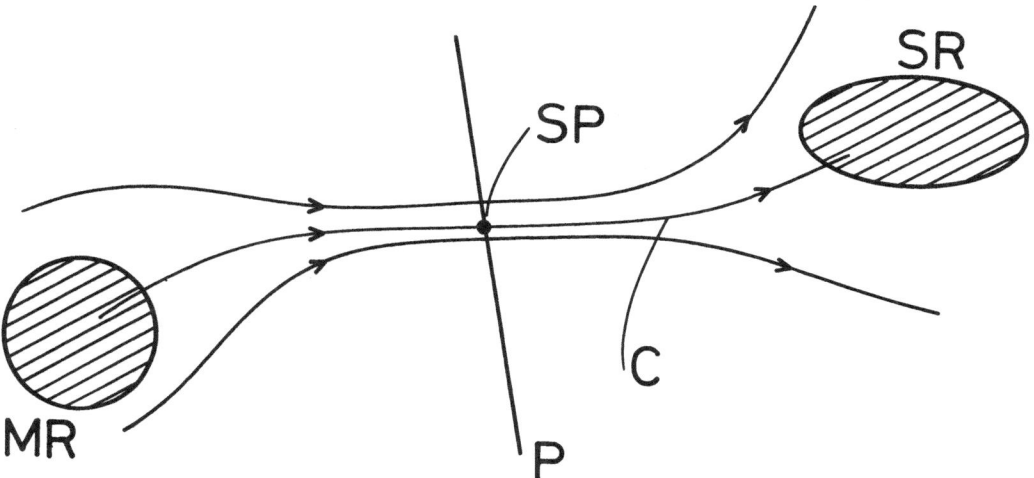

<u>Figure 6.1</u> : Schematic illustration of the probability current flowing
from a metastable state (MR is the metastable region)
to a stable state (SR is the stable region) across a
saddle point SP. P is the projection of the hyperplane
u = 0.

This involves a consideration of the static case. The diagonaliza-
tion of the quadratic form $\mathbf{M}$ defined a local coordinate system around
the saddle point. Indeed, we can write for $c(\vec{x})$ in the vicinity of
the saddle point :

$$c(\vec{x}) = m(\vec{R}) + \sum_{\ell} \eta_{\ell} \varphi_{\ell}(\vec{x}) . \qquad (6.3)$$

In the local coordinate system, the components of the probability current are :

$$J_{\ell} = \int d^3x \, \varphi_{\ell}(\vec{x}) J(\vec{x}) = - \sum_{\ell'} \bar{\Gamma}_{\ell\ell'} \cdot [\frac{\partial F}{\partial \eta_{\ell'}} P + \frac{\partial P}{\partial \eta_{\ell'}}] \qquad (6.4)$$

with $\bar{\Gamma}_{\ell\ell'} = \int d^3x \, (\vec{\nabla}\varphi_{\ell}) \Gamma (\vec{\nabla}\varphi_{\ell'})$. $\qquad (6.5)$

The condition for the steady state can then be written as :

$$\sum_\ell \frac{\partial J_\ell}{\partial n_\ell} = - \sum_{\ell\ell'} \frac{\partial}{\partial n_\ell} \bar{\Gamma}_{\ell\ell'} [\frac{\partial F}{\partial n_{\ell'}} P_s + \frac{\partial P_s}{\partial n_{\ell'}}] = 0 \tag{6.6}$$

where P_s denotes the steady state solution. For the physical reasons discussed above, it is reasonable to look for a steady state solution of the form :

$$P_s(\eta) = Q(\eta) \exp - (F(\eta)) . \tag{6.7}$$

Using moreover the fact that, close to the saddle point [*], the coarse-grained free energy is (see Eq. (5.60)) :

$$F(\eta) = F_{dr} + \frac{1}{2} \sum_\ell \omega_\ell n_\ell^2 \tag{6.8}$$

the stationary condition for $Q(\eta)$ becomes :

$$\sum_{\ell\ell'} \bar{\Gamma}_{\ell\ell'} [\frac{\partial^2 Q}{\partial n_\ell \partial n_{\ell'}} - \omega_\ell n_\ell \frac{\partial Q}{\partial n_{\ell'}}] = 0 . \tag{6.9}$$

The solution of this equation is of the form :

$$Q(\eta) = f(u) \tag{6.10}$$

where u is a linear combination of the n_ℓ, excluding n_o. Thus :

$$u = \sum_\ell{}' u_\ell n_\ell \tag{6.11}$$

the prime excluding $\ell = 0$ from the sum. Substituting into (6.9) one obtains the following differential equation for f :

$$(\sum_{\ell\ell'}{}' u_\ell \bar{\Gamma}_{\ell\ell'} u_{\ell'}) \frac{d^2 f(u)}{du^2} - (\sum_{\ell\ell'}{}' \omega_\ell n_\ell \cdot \bar{\Gamma}_{\ell\ell'} u_{\ell'}) \frac{df(u)}{du} = 0 . \tag{6.12}$$

This equation makes sense (differential equation of the variable u only) if the coefficient of $df(u)/du$ is proportional to u itself. Thus :

---

[*] Note that in order to simplify the notation the index ℓ stands for (n,ℓ) used in the preceding section.

$$\sum_{\ell\ell'}' \omega_\ell \eta_\ell \bar{\Gamma}_{\ell\ell'} u_{\ell'} = Ku = K \sum_\ell' u_\ell \eta_\ell \ . \tag{6.13}$$

This gives a condition for the coefficients u_ℓ which should be a solution of the eigenvalue equation

$$\omega_\ell \sum_{\ell'}' \bar{\Gamma}_{\ell\ell'} u_{\ell'} = Ku_\ell \ . \tag{6.14}$$

With $c = \dfrac{1}{K} \sum_{\ell\ell'}' u_\ell \bar{\Gamma}_{\ell\ell'} u_{\ell'} \ ,$ $\hspace{3cm}$ (6.15)

equation (6.12) becomes :

$$c\frac{d^2f}{du^2} - u\frac{df}{du} = 0 \ . \tag{6.16}$$

If c is positive, the only solution is $f(u) = $ cte. However, for c negative, f is proportional to the error function (see Eq. (4.27)) and:

$$f(u) = f_o \mathrm{erf}(u/\sqrt{2}|c|) \sim \int du \ \exp\ (\frac{u^2}{2c}) \ . \tag{6.17}$$

Let us now return to the eigenvalue equation (6.13) and show that this equation has a very simple physical interpretation. To see that, let us compute the average of $<\eta_\ell(t)>$ where

$$<\eta_\ell(t)> \equiv \int D(\eta) \eta_\ell P[\{\eta\},t] \ . \tag{6.18}$$

In terms of the (η) variables, the Fokker-Planck equation (6.1) becomes :

$$\partial_t P[\{\eta\},t] = -\sum_\ell \frac{\partial J_\ell}{\partial \eta_\ell} \ . \tag{6.19}$$

Thus $\partial_t <\eta_\ell(t)> = -\sum_\ell \int D\{\eta\} \eta_\ell \frac{\partial J_\ell}{\partial \eta_\ell} = \int D(\eta) J_\ell$ $\hspace{1.5cm}$ (6.20)

the last result being obtained by partial integration. Moreover from (6.4) and (6.6), it follows that

$$\int D\{\eta\} J_\ell = -\sum_{\ell'} \bar{\Gamma}_{\ell\ell'} \omega_\ell <\eta_{\ell'}(t)> \ . \tag{6.21}$$

Thus $\{<n_\ell(t)>\}$ satisfies the simplest linear differential equation :

$$\partial_t <n_\ell(t)> = -\sum_{\ell'} \bar{\Gamma}_{\ell\ell'}\omega_\ell <n_\ell(t)> . \tag{6.22}$$

Note that the linearity of this equation is due to the quadratic approximation used for the free energy around the saddle point. Now, if as we have assumed the saddle point is to describe the nucleating fluctuation, then there must be one particular direction around the saddle point in which the solution of (6.22) is unstable. Thus there must be one solution of the form

$$<n_\ell(t)> = A_\ell \exp - Kt \quad \text{with} \quad K < 0 . \tag{6.23}$$

Thus K is just the initial growth rate of the unstable mode at the saddle point. One assumes that this unstable mode is unique. Since K is negative, it turns out that c defined by equation (6.15) is also negative and then (6.17) is the good solution for $f(u)$.

From (6.22), we see that the condition $u = 0$ defines an hyperplane passing through the saddle point and not tangent to the direction n_o. Moreover, one assumes that the two relevant stable and metastable minima of $F(\eta)$ lie on opposite sides of the hyperplane $u = 0$ and are well out of the transition region. Accordingly, one chooses $f(u)$ such that $P_s \simeq P_o$ finite in the metastable region and $P_s \simeq 0$ in the stable one. One aims to find the probability current flowing from the metastable to the stable region. There is a source of current on the metastable side and a sink on the stable side (see Fig. 6.1). The convention adopted is that metastable configurations occur at negative values of u. Thus the above considerations and the result (6.17) lead one to choose

$$f(u) = \frac{1}{z_o \sqrt{2\pi|c|}} \int_u^\infty du' \exp (-\frac{u'^2}{2|c|}) \tag{6.24}$$

z_o is chosen so that P_s is normalized in the metastable phase. Thus

$$1 = \int\limits_{(u<0)} D\{n\}P_s\{n\} = \frac{1}{z_o} \int\limits_{(u<0)} D\{n\}\exp(-F\{n\}) \frac{1}{\sqrt{2\pi|c|}} \int_{-\infty}^\infty du' \exp -\frac{u'^2}{2|c|}$$

$$\tag{6.25}$$

and thus $\quad Z_o = \int\limits_{(u<0)} D\{\eta\} \exp(-F\{\eta\})$ \hfill (6.26)

Z_o is the partition function computed in the preceeding chapter whose value is given by Eq. (5.58).

Let us now calculate the probability current flowing across the saddle point. Using the above intermediate results, we can rewrite equation (6.4) as :

$$J_\ell = -\sum_{\ell'} \bar{\Gamma}_{\ell\ell'} \, \frac{\partial Q}{\partial \eta_{\ell'}}\{\exp{-F}\} = -\sum_{\ell'} \bar{\Gamma}_{\ell\ell'} u_{\ell'} \, \frac{df}{du}\{\exp{-F}\} \qquad (6.27)$$

using (6.14) and (6.17) one obtains :

$$J_\ell = +\exp{-F}\{\eta\}[\frac{1}{Z_o\sqrt{2\pi|c|}}]K \frac{u_\ell}{\omega_\ell} \exp{-\frac{1}{2|c|}} \sum_{\ell\ell'}' u_\ell u_{\ell'} \eta_\ell \eta_{\ell'} \qquad (6.28)$$

finally using (6.8), one gets :

$$J_\ell = \exp{-F_{dr}} \frac{K}{Z_o\sqrt{2\pi|c|}} \frac{u_\ell}{\omega_\ell} \exp{[-\frac{1}{2}\sum_\ell \omega_\ell \eta_\ell^2 - \frac{1}{2|c|}} \cdot$$

$$\cdot \sum_{\ell\ell'}' u_\ell u_{\ell'} \eta_\ell \eta_{\ell'}] \quad . \qquad (6.29)$$

Thus close to the saddle point, the probability current has a constant direction parallel to the vector whose components are u_ℓ/ω_ℓ. To obtain the nucleation rate I, we have to compute the flux of probability current across a surface passing through the saddle point and not parallel to the current. Several choices are possible. A suitable choice is to integrate over the surface defined by $\eta_1 = 0$. Indeed in this case the nucleation rate is given by

$$I = \int\limits_{(\eta_1=0)} J_1\{\eta\}d\sigma_1 d\eta_1 \qquad (6.30)$$

where $d\sigma_1$ is a surface element of $\{\eta_1 = 0\}$. Thus one needs to perform Gaussian integrals over the modes $\eta_\ell \neq \eta_1$. The contribution due to η_1 is proportional to the volume of the plane $(\eta_1 = 0)$ in the configuration space. Thus this is exactly the calculation that we did to compute the imaginary part of the free energy in Chapter 5. Thus, the final result for the nucleation rate is :

$$I = \frac{|K|}{\pi} \, \text{Im} f(h) \tag{6.31}$$

Note that (6.31) gives a physical significance to $\text{Im} f(h)$ which is not apparent in the discussion of Chapter 5.

Using (5.105) for the imaginary part of the free energy, we re-write I as :

$$I = I_o e^{-\Delta F} \tag{6.32}$$

with $I_o = \frac{|K|}{2\pi} \, \Omega_o$, \hfill (6.33)

where ΔF is the activation energy given by (5.71). Within the (more explicit) notation of Chapter 5, Ω_o is :

$$\Omega_o = \Gamma_3 \left(\frac{2\pi}{|\omega_{oo}|}\right)^{1/2} \prod_{n,\ell}{}' \left(\frac{2\pi}{\omega_{n,\ell}}\right)^{(2\ell+1)/2} \prod_j \left(\frac{\omega_j^o}{2\pi}\right)^{1/2} \tag{6.34}$$

Γ_3 is given by (5.94) and is proportional to the volume V of the system. (Since I is proportional to the volume of the system, it is obvious that Ω_o must contribute such a V dependence.) Ω_o is called the statistical factor while $|K|$ is called the dynamical factor. Thus any nucleation rate calculation within this formalism requires the computation of three quantities : The activation energy of the critical droplet ΔF, the statistical factor Ω_o linked to the eigenvalue spectrum $\{\omega_\ell\}$ and $\{w_j^o\}$ and finally the dynamical factor $|K|$. This factor can be obtained by linearizing the equation of motion (non-linear Langevin or Fokker-Planck equations) around the saddle point.

We see that this field theoretical approach has some similarities with the Becker-Döring theory. Both assume that the metastable states decay via thermal activation of localized unstable fluctuations. However, in the field theoretical approach, one does not need a detailled theory describing the formation of the droplets. Moreover the field theoretical approach allows one to take significant fluctuations into account.

It should be noted that a significant advantage of the field theoretic approach over the original classical droplet model calculations is that it starts with a first principles, statistical description of all relevant degrees of freedom as modelled by the "Hamiltonian" (5.1). It then treats the "droplets" as a single collective mode, which is given by the solution of the saddle point equation. Fluctuations are taken into account by an expansion around the critical droplet solution, which corresponds to the "surface excitations" of the droplet. Finally, it is never assumed that the "critical droplet" corresponds to a physical droplet. The latter identification historically has led to great difficulties due to the problems inherent in giving a precise definition of a droplet. In the Langer approach, droplets only are defined via the saddle point solution.

REFERENCES - Chapter 6.

[1] J.S. Langer, Ann. Phys. (N.Y.) 54, 258 (1969).

[2] R. Landauer and J.A. Swanson, Phys. Rev. 121, 1668 (1961).

GENERAL REFERENCES

[G1] J.S. Langer and L.A. Turski, Phys. Rev. A8, 3230 (1973), Phys. Rev.
A22, 2189 (1980). These papers apply the general formalism of this
chapter to homogeneous nucleation near the critical point of simple and binary fluids.

[G2] J.S. Langer, in "Systems Far From Equilibrium", Lecture Notes in
Physics 132, L. Garrido ed., Springer-Verlag (1980). This article
compares the theoretical predictions of the above papers with recent precision experiments on fluids.

[G3] J.S. Langer and A.J. Schwartz, Phys. Rev. A21, 948 (1980). This
paper is the first attempt to give a full description of the phase
separation process. This includes a theoretical treatment of the
very difficult problem posed by the simultaneous occurence of the
birth and growth of droplets. This problem warrants further theoretical investigation.

[G4] W.I. Goldburg, in "Scattering Techniques Applied to Supramolecular
and Nonequilibrium Systems", edited by S.M. Chan, B. Chu and
R. Nossal, Plenum Press (New York), (1981). This includes a nice
discussion of experiments on nucleation in fluids, as well as an
extensive list of references to the experimental literature.

[G5] R.G. Howland, N.C. Wong and C.M. Knobler, J. Chem. Phys. 73, 522
(1980). This is a very nice study of homogeneous nucleation in
near-critical fluids, including a careful discussion of various
experimental techniques and difficulties.

CHAPTER 7.   THEORIES OF SPINODAL DECOMPOSITION

A major unsolved problem in the dynamics of first order phase transitions is the dynamical behavior of a system following a quench into the unstable region of the phase diagram. In this chapter we summarize several attempts to understand the early stages of this instability which is often termed spinodal decomposition. These include linear theories due primarily to Hillert [1], Cahn [2] and Cook [3] and the most successful non-linear theory so far developed, due to Langer, Bar-on and Miller [4]. In this chapter most of the details are omitted, because there is at the moment no completely satisfactory theory.

7.1   Linear Theories

Our starting point is the equation of motion (3.38) for the structure factor $S(\vec{k},t)$ which can be written as :

$$\partial_t S(\vec{k},t) = -2Mk^2[(Ck^2 + V_o^{(2)})S(\vec{k},t) + \sum_{n=3}^{\infty} \frac{1}{(n-1)!} V_o^{(n)} S_n(\vec{k},t)]$$

$$+ 2Mk_B T k^2 . \tag{7.1}$$

The first qualitative theoretical understanding of the long-wavelength instability which characterizes spinodal decomposition is due to Cahn. His theory is a deterministic theory in which the noise term in the Langevin equation (3.30) or equivalently $2Mk_B T k^2$ in (7.1) is neglected. Cahn noted that immediately following a quench into the unstable region the initial fluctuation in concentration should be small, so that one should be able to linearize the nonlinear Langevin equation (3.28), or equivalently, neglect all the higher order correlation functions in the deterministic version of (7.1). This leads to the linear equation

$$\partial_t S(\vec{k},t) \cong -2Mk^2[(Ck^2 + V_o^{(2)})]S(\vec{k},t) \tag{7.2}$$

whose solution is

$$S(\vec{k},t) = S(\vec{k},0)e^{-w(k)t} \tag{7.3}$$

where

$$w(k) = 2MCk^2 [k^2 - k_c^2] \qquad (7.4)$$

and

$$k_c^2 = \frac{1}{C} v_o^{(2)} \quad . \qquad (7.5)$$

(Recall that in the unstable region $v_o^{(2)} < 0$ by definition.) Thus according to the Cahn theory the structure factor should initially grow exponentially with time for all $k < k_c$, since $w(k) < 0$ for $k < k_c$. For systems for which multiple scattering can be neglected, the intensity of scattered radiation $I(\vec{k},t)$ is proportional to $S(\vec{k},t)$. Thus the initial long wavelength instability should be manifest as an exponential growth in $I(\vec{k},t)$, with a time independent maximum at $k = k_m = k_c/\sqrt{2}$. This exponential growth should correspond to the appearance of a fine, uniformly dispersed precipitate whose pattern should exhibit a near-periodicity at a wavelength $\lambda_m = 2\pi/k_m$. Although original experiments on AlZn alloys were interpreted as verifying (7.3), later studies showed this not to be the case (see for example the reviews by Gunton, San Miguel and Sahni [5] and by Gerold and Kostorz [6]).

An important refinement of the Cahn theory was made by Cook, who proposed that a noise term should be added to the deterministic equation studied by Cahn. Cook's equation is model B, equation (3.28). The Cook theory corresponds to keeping the term $2Mk_BT\,k^2$ in (7.1) but again neglecting the highest order correlation functions $S_n(\vec{k})$. The resulting inhomogeneous equation can be solved in the usual way, using the homogeneous solution (7.3). The result is

$$S(\vec{k},t) = S(\vec{k},0)e^{-w(k)t} + S(\vec{k},\infty)[1 - e^{-w(k)t}] \quad . \qquad (7.6)$$

Here $S(\vec{k},\infty)$ denotes the "equilibrium" value of the structure factor at the final quench temperature, for wavelengths $k > k_c$. The Cook theory has the merit that it correctly describes the relaxation of fluctuations with wave numbers $k > k_c$ to an Ornstein-Zernike type "equilibrium" [7]. In contrast, the Cahn theory predicts that the structure factor decays to zero for $k > k_c$, which is incorrect. The major advance made by Cook, however, was to introduce into the metallurgical literature the concept that a stochastic theory is necessary for a proper description of phase separation. His linear theory, (7.6), however, does not agree in general with the experimental measurements. (A few exceptions have been claimed in the literature.) It is clearly necessary, therefore, to take into account nonlinear terms to adequately explain the experimental and Monte Carlo results. (A discussion of the expected domain of validity of the linear theory is given by Skripov and Skripov [G3].)

7.2 The Langer, Bar-on, Miller Theory

The only reasonably successful early time theory of spinodal decomposition which treats the nonlinear dynamical effects is due to Langer, Bar-on and Miller [4]. Their theory is a particular truncation of the exact model equation of motion (7.1) which is plausible but probably not systematic. They assume that the higher order correlation functions $S_n(\vec{k},t)$ can be approximated by :

$$S_n(\vec{k},t) \cong \frac{\langle u^n \rangle}{\langle u^2 \rangle} S(\vec{k},t) \tag{7.7}$$

where $u(\vec{x},t)$ is defined in (3.32) and

$$\langle u^n(\vec{x},t) \rangle = \int D[u] u^n(\vec{x}) P(u(\vec{x}),t) \tag{7.8}$$

$$= \int du \, u^n(\vec{x}) P_1(u(\vec{x}),t) \ . \tag{7.9}$$

The one point distribution functional P_1 in (7.9) is given by the m-point distribution functional P_m with m = 1, where P_m is defined as

$$P_m(u(\vec{x}_1),u(\vec{x}_2),\ldots,u(\vec{x}_m);t) = \int^{(m)} D[u] P(u(\vec{x}),t) \ . \tag{7.10}$$

The functional integral on the right-hand side of (7.10) denotes integrating over all u space while holding the values of u fixed at points $\vec{x}_1,\ldots,\vec{x}_m$. Before we discuss the origin of (7.7), we note that with this approximation (7.1) reduces to the much simpler equation

$$\partial_t S(\vec{k},t) = -2Mk^2(Ck^2 + A(t))S(\vec{k},t) + 2Mk_B T \, k^2 \tag{7.11}$$

where

$$A(t) = \sum_{n=2}^{\infty} \frac{1}{(n-1)!} V_o^{(n)} \frac{\langle u^n \rangle}{\langle u^2 \rangle} \ . \tag{7.12}$$

To determine $A(t)$ requires knowing $\langle u^n \rangle$ which is obtained from P_1 and (7.9). The one-point distribution P_1 is, however, determined from the ansatz for P_2 which leads to (7.7), so that in principle (7.10) with (7.11), determine $S(\vec{k},t)$.

To obtain (7.7) we first note that by the definition (3.33) $S_n(\vec{x}-\vec{x}',t)$ only requires knowledge of P_2, since

$$S_n(\vec{x}-\vec{x}',t) = \int^{(2)} D[u] u^{(n-1)}(\vec{x}) u(\vec{x}') P_2(u(\vec{x}),u(\vec{x}'),t) \ . \tag{7.13}$$

Langer, Bar-on and Miller (LBM) assume that the quantity

$$P_2(u(\vec{x}),u(\vec{x}')) - P_1(u(\vec{x})) P_1(u(\vec{x}'))$$

can be expanded in a functional Taylor series expansion in the fluctuations $u(\vec{x})$ and $u(\vec{x}')$. They then truncate this expansion at the lowest order term, which leads to

$$P_2(u(\vec{x}),u(\vec{x}')) = P_1(u(\vec{x})) P_1(u(\vec{x}')) [1 + f(|\vec{x}-\vec{x}'|) u(\vec{x}) u(\vec{x}') + \ldots] \tag{7.14}$$

The one point function P_1 satisfies the conditions that

$$\int_{-\infty}^{\infty} P_1(u) du = 1 \tag{7.15}$$

$$\int_{-\infty}^{\infty} P_1(u) u \, du = 0 \tag{7.16}$$

where (7.16) is obviously true by the definition of u. The normalization of P_2 immediately follows, since from (7.14), (7.15) and (7.16)

$$\int_{-\infty}^{\infty} P_2(u(\vec{x}),u(\vec{x}')) du(\vec{x}) du(\vec{x}') = 1 \ . \tag{7.17}$$

In addition, it follows from the definition of $S(\vec{x}-\vec{x}')$ (Eq. (7.13) with $n = 2$) and (7.14) that

$$S(|\vec{x}-\vec{x}'|) = f(|\vec{x}-\vec{x}'|)<u^2>^2 \quad . \tag{7.18}$$

This identifies the function $f(|\vec{x}-\vec{x}'|)$ in (7.14). One then obtains from (7.13) and (7.14) the relation

$$S_n(\vec{r}) \simeq \frac{<u^n>}{<u^2>} S(\vec{r}) \tag{7.19}$$

whose Fourier transform is (7.7).

It is difficult to justify the validity of (7.14) or (7.19). For large spatial separations ($|\vec{x}-\vec{x}'|$ large) (7.19) seems reasonable, but the only test of the approximation is its a posteriori comparison with experiment. The best comparison which has been made is with Monte Carlo simulations of the kinetic Ising model of a binary alloy, which can be thought of as "numerical experiments".

To proceed further with the theory, we need two additional quantities. The first is the coarse-grained free energy function $V(c)$ which appears in (7.1) and (7.12). This quantity has not yet been computed from first principles, as noted in Section 2.3. Langer et al., assume that for a coarse-graining size L which is proportional to the correlation length ξ,

$$V(c) \simeq \frac{k_B T}{L^3} f_0 [-\frac{1}{2}(\frac{c}{c_s})^2 + \frac{1}{4}(\frac{c}{c_s})^4] \quad , \tag{7.20}$$

where f_0 is a normalization constant and $c_s(T)$ is the equilibrium coexistence value of the order parameter (the miscibility gap for a binary alloy). Near the critical point the asymptotic behavior of c_s is [*] :

---

[*] The exponents β, γ, ν, η introduced below are the usual critical exponents describing the behaviour of certain physical quantities close to a critical point. The reader not familiar with those concepts can refer to the introduction of Stanley [8].

$$c_s \simeq \pm B\epsilon^\beta, \quad \epsilon = 1 - T/T_c \ . \tag{7.21}$$

It should be noted that although $V(c)$ is a nonconvex function, Langer et al. assume that it is closely related to the equilibrium free energy, for $L \simeq \xi$ since V then contains most of the thermodynamically important fluctuations. For example, they assume that $V(c)$ has its minima at the equilibrium values of the order parameter, $c = \pm c_s$. This is almost certainly not strictly true, but the error involved could be small. Likewise they assume that f_o can be determined in terms of known equilibrium properties of the system. Their argument is that the static susceptibility

$$\chi = \frac{1}{a_o^3} S(k=0) \simeq E\epsilon^{-\gamma} \tag{7.22}$$

(where a_o is the lattice spacing) can be obtained from $V(c)$ via the estimate

$$S(k=0) \simeq k_B T/(\partial^2 V/\partial c^2)_{c_s} \tag{7.23}$$

$$= L^3 c_s^2/2f_o \ . \tag{7.24}$$

Again, (7.23) is not exact, since $S(0)$ and χ are determined from the convex, equilibrium free energy, rather than from the nonconvex $V(c)$. However, using this approximation one can determine f_o, since from (7.21), (7.22) and (7.24)

$$f_o = (\frac{L}{a_o})^3 \frac{c_s^2}{2\chi} \simeq (\frac{L}{a_o})^3 \frac{B^2}{2E} \epsilon^{2\beta+\gamma} \tag{7.25}$$

Langer et al. then choose

$$L = \xi/\alpha \tag{7.26}$$

with $\alpha = (6\pi^2)^{-1/3}$, which leads to

$$f_o \simeq 6\pi^2 (\frac{\xi_o}{a_o})^3 \frac{B^2}{2E} \tag{7.27}$$

where

$$\xi \cong \xi_o \epsilon^{-\nu} \tag{7.28}$$

since $2\beta + \gamma = 3\nu$ is a hyperscaling relation [8]. Thus f_o can be estimated from the amplitudes ξ_o, B and E, all of which are accurately known for the three dimensional Ising model. (It should be noted that the above procedure for determining f_o has been criticized by Billotet and Binder [9] as being inconsistent, in that it leads to an incorrect value for the critical amplitude of the susceptibility and static structure factor.) The second quantity which one needs to estimate is the dynamical coefficient, M, which appears in (7.1). Langer, Bar-on and Miller assume that for the kinetic Ising model M depends only upon the rate of attempted atomic exchanges and the underlying lattice structure, but not on dynamical quantities such as interaction energies. This leads them to estimate that for a simple cubic Ising model

$$M \simeq (\frac{a_o^5}{12k_B T})(1 - c^2) . \tag{7.29}$$

This c-dependence of M is in disagreement with the original Fokker-Planck model, in which M is taken to be independent of c. The authors replace c in (7.29) by c_s, however, so that at this level no contradiction exists.

The calculation of $S(k,t)$ from (7.11) and (7.12) is carried out numerically. It is natural to introduce a scaled time, wavenumber, and structure factor via

$$q = \sqrt{2}k\xi \tag{7.30}$$

$$\tau = A_1 \xi^{-z} t \tag{7.31}$$

and

$$\mathcal{S}(q,\tau) = A_2 \xi^{-\gamma/\nu} S(k,t) , \tag{7.32}$$

where $A_1 = Mk_B T/(2a_o^3 \xi_o^{2-z} E)$, $A_2 = \xi_o^{\gamma/\nu}/2Ea_o^3$ and $z = \gamma/\nu + 2 = 4 - \eta$, where η is the exponent for the order parameter correlation function

(i.e., $S(k,T = T_c) \sim k^{-(2-\eta)}$). The scaled forms of (7.11) and (7.5)
are

$$\partial_\tau \mathcal{S} = -q^2(q^2-\mu)\mathcal{S} + q^2 \qquad (7.33)$$

and

$$\mu = -\frac{1}{\langle y^2 \rangle} \langle y \frac{\partial \phi(x_0+y)}{\partial y} \rangle . \qquad (7.34)$$

Here

$$y = u/c_s , \quad x_0 = c_0/c_s \qquad (7.35)$$

and

$$\phi(y) = -\frac{1}{2}y^2 + \frac{1}{4}y^4 . \qquad (7.36)$$

Thus within the context of the Ginzburg-Landau free energy approximation, (7.20), the scaled variable $\mu \langle y^2 \rangle$ is just a fourth-order polynomial. For example, for a critical quench composition $c_0 = 0$ for the
Ising model,

$$\mu = -\langle y^2 \rangle^{-1} [-\langle y^2 \rangle + \langle y^4 \rangle] . \qquad (7.37)$$

The solution of (7.33) and (7.34) requires numerically determining the
scaled one-point distribution function $P_1(y,\tau)$ in order to determine
$\mu(\tau)$ in (7.34). An equation of motion is obtained for $P_1(\tau)$ from the
original Fokker-Planck equation (3.32), using the approximation (7.14).
The authors parametrize P_1 as the sum of two Gaussians, with three
parameters which determine the two peak positions and half width respectively. As this calculation is complicated, we do not reproduce it
here. The results of such a solution of (7.33) and (7.34) are shown in
Fig. 7.1 and Fig. 7.2 for a symmetric quench $c_0 = x_0 = 0$ and asymmetric
quench $(x_0 = c_0/c_3 = 1/\sqrt{3})$ respectively. (The asymmetric quench corresponds to a point on the classical spinodal curve, where the linear
theory would incorrectly predict no phase separation since the driving
force is zero there.) As can be seen, the theory quite satisfactorily
explains the main qualitative features of the early time development
for both quenches.

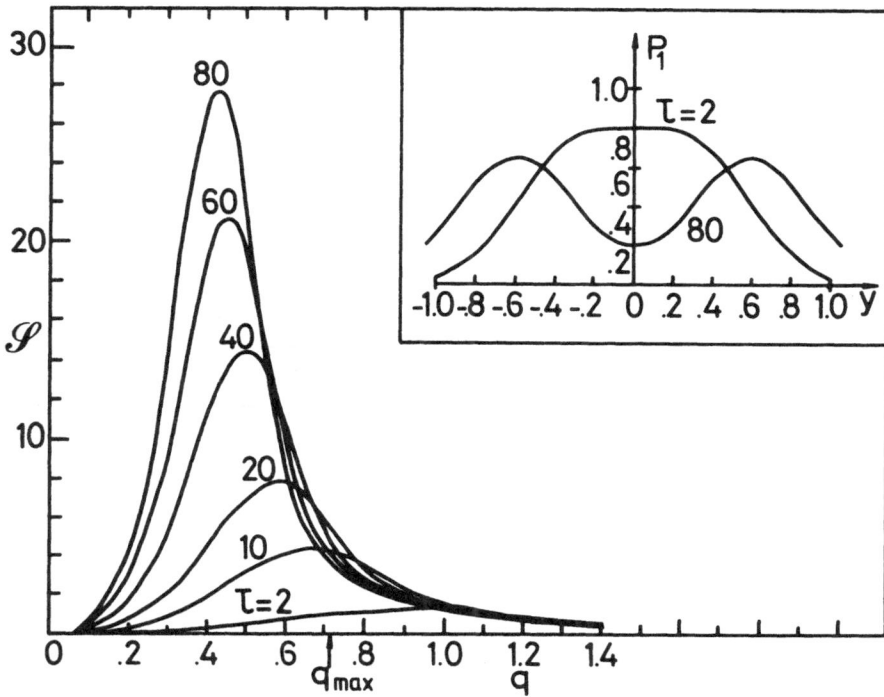

Figure 7.1 : The scaled structure factor $\mathcal{S}(q)$ for a quench at critical composition at various scaled times τ. The insert
depicts the distribution function $P_1(y)$ at two of these
times. (From Langer, Bar-on and Miller [4].)

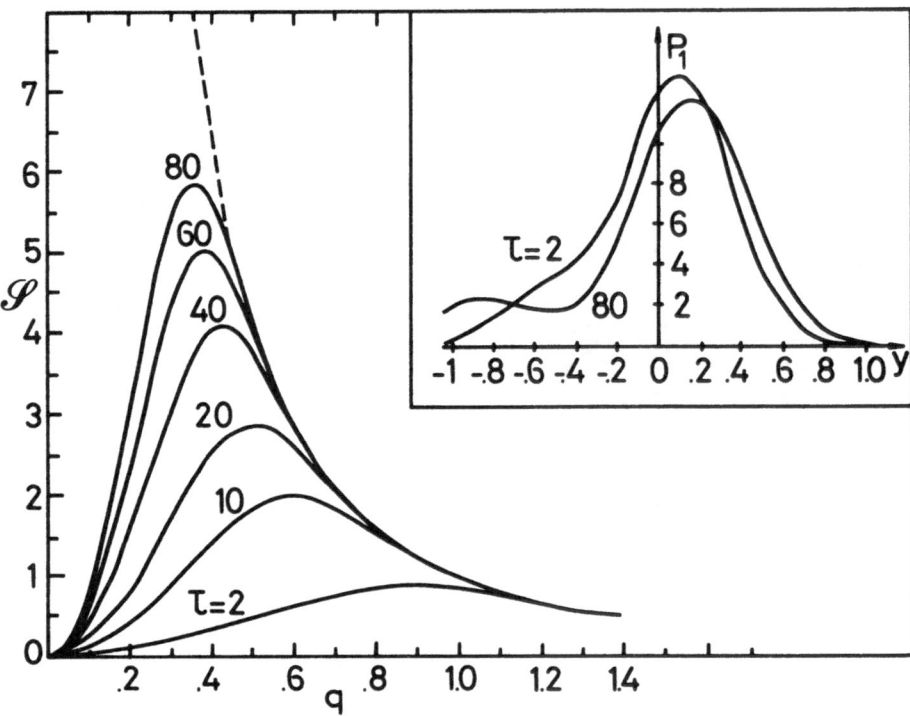

Figure 7.2 : The scaled structure factor $\mathcal{S}(q)$ for the asymmetric
quench described in the text. The insert depicts the distribution function $P_1(y)$, at two of these times. (From
Langer, Bar-on and Miller [4].)

Not only does a peak in $\mathcal{S}(q,\tau)$ develop and increase with time, but "coarsening" is manifest in the fact that the position of the peak, $q_m(\tau)$, decreases with increasing τ. Evidence for true phase separation is shown by the development of two distinct peaks in P_1. Langer, Bar-on and Miller compared their theory with results from the Monte Carlo study of the kinetic Ising models [10], for a critical composition quench. The agreement is quite reasonable for early times, particularly when one realizes that there are no adjustable parameters. The theory, however, becomes inaccurate for later times. In particular, it does not describe the late stage approach to equilibrium correctly. Although the results for the asymmetric quench are qualitatively correct, the agreement with Monte Carlo results is less satisfactory than for the critical quench.

In spite of the success of this theory, which is noteworthy in comparison with earlier work, the theory has several limitations which have been pointed out by Billotet and Binder [9] and Binder, Billotet and Mirold [11]. We summarize these weaknesses here, since they clarify the direction for future research in this area. One difficulty has to do with the original conclusion that the theory describes reasonably well the expected gradual transition from spinodal decomposition to nucleation as one varies the quench concentration (e.g., Fig. 7.2). Billotet and Binder noted that the LBM theory actually is incapable of describing nucleation and growth. This shows up in the fact that there are one-phase state, stationary solutions of the LBM equations (7.33) and (7.34) in the interior of the coexistence curve, between the coexistence curve and a pseudo spinodal curve which are the final equilibrium solutions of the LBM equations, (the precise location of this pseudo-spinodal depends on the parameter α in (7.27)). These stationary solutions describe metastable states with <u>infinite</u> lifetime. The LBM theory thus yields a structure factor in this region which approaches an Ornstein-Zernike-like form appropriate for such metastable states, as shown in Fig. 7.3. This behaviour does not describe the nucleation and growth regime, so that the LBM theory becomes qualitatively incorrect as one quenches near the coexistence curve. A second difficulty, which is in retrospect not surprising given the LBM truncation of the equations for $S(\vec{k},t)$, is that certain results of the LBM theory depend rather sensitively on the parameter α in (7.26), which defines the coarse-graining size. Since this coarse-graining size should be an irrelevant parameter, this is an incorrect feature of the theory. Perhaps the most

striking manifestation of this dependence is seen in the continuation
of the equilibrium solution of the LBM equations in the one-phase region into the metastable region.

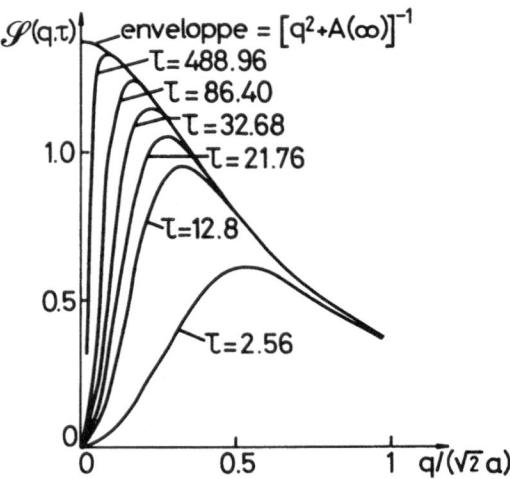

Figure 7.3 : Time-evolution of the scaled structure factor $S(q,\tau)$ for
a quench near the coexistence curve as calculated from the
equation of Langer, Bar-on and Miller [4] by Binder,
Billotet and Mirold [11].

One finds that there are different "metastable" branches for different
values of α. (It should also be pointed out that the mean field theory for such a continuation corresponds to the choice $\alpha = 0$.) There
is also a pseudo-spinodal in the LBM theory, which is the end-point of
the metastable branch. The numerical results of the Billotet-Binder
analysis suggest that at this point the static susceptibility χ
(given by (7.22)) remains finite at this pseudo-spinodal, but its derivative, $\partial\chi/\partial h$, diverges there. The precise location of such a pseudo-
spinodal depends on the choice of α. Billotet and Binder have also
shown that a certain nonlinear relaxation time diverges at this pseudo-
spinodal, whereas in fact it should diverge only at the coexistence
curve.

To summarize, the LBM theory represents a step forward in our understanding of the early stages of spinodal decomposition for near-critical quenches. However, it becomes inaccurate at later times and for quenches near the coexistence curve. There is a clear need to develop the theory so as to incorporate nucleation and growth as well, but this is a nontrivial problem. The dependence of some of the results of the theory on the coarse-graining parameter α emphasizes the need to obtain renormalized equations of motion in which such a parameter drops out. This seems to be a problem for which renormalization group methods would be very useful. Apart from one attempt by Horner and Jungling [12] (which in our opinion is not a systematic analysis) to perform a first order in $\epsilon = 4 - d$ renormalization, there has been no progress on this problem.

REFERENCES - Chapter 7.

[1] M. Hillert, Acta. Metall. 9, 525 (1961).

[2] J.W. Cahn, Trans. Metall. Soc. AIME 242, 166 (1968) and references contained therein to his original work from 1961 to 1968.

[3] H.E. Cook, Acta. Metall. 18, 297 (1970).

[4] J.S. Langer, M. Bar-on and H.D. Miller, Phys. Rev. A11, 1417 (1975).

[5] J.D. Gunton, M. San Miguel and P.S. Sahni, to be published in Vol. 8, "Phase Transitions and Critical Phenomena", Academic Press (London), edited by C. Domb and J.L. Lebowitz (1983).

[6] V. Gerold and G. Kostorz, J. Appl. Cryst. 11, 375 (1978).

[7] L.S. Ornstein and F. Zernicke, Proc. Sect. Sci. K. Med. Akad. Wet. 17, 793 (1914).

[8] H.E. Stanley, "Introduction to Phase Transitions and Critical Phenomena", Ed. Clarendon Press Oxford (1971).

[9] C. Billotet and K. Binder, Zeit. Physik B32, 195 (1978).

[10] J. Marro, A.B. Bortz, M.H. Kalos and J.L. Lebowitz, Phys. Rev. B12, 2000 (1975).

[11] K. Binder, K. Billotet and P. Mirold, Zeit. Physik B30, 183 (1978).

[12] H. Horner and K. Jüngling, Zeit. Physik B36, 97 (1979).

GENERAL REFERENCES

[G1] E.D. Siggia, Phys. Rev. A20, 595 (1979). This contains an interesting discussion of several late stage growth mechanisms in fluid mixture, where hydrodynamics can play an important role.

[G2] K. Kawasaki and T. Ohta, Prog. of Theor. Phys. 67, 147 (1982).

This paper seems to provide a promising starting point for a successful theory of interface dynamics and the corresponding structure factor.

[G3] V.P. Skripov and A.V. Skripov, Sov. Phys. Uspkeki 22, 389 (1979). This is a detailed review of the general phenomena of spinodal decomposition, including many aspects not discussed in this chapter.

8.1 Nonlinear Equations of Motion

One of the few (if not the only) reasonably well established theories which deal with dynamical phenomena in first order phase transitions is the Lifshitz-Slyozov (LS) theory for the late stage growth of droplets [*]. This theory deals with the case of small initial supersaturation, in the asymptotic time domain ($t \to \infty$). For such small supersaturations and at such late times, the birth of new droplets (nucleation) is negligible, due to the large critical sizes involved. In this late time domain one wishes to describe how the droplets of the minority phase subsequently evolve to attain the final equilibrium state. We will consider the case of a solid solution in which the droplets are at rest and will assume for simplicity that the droplets (grains) are spherical and that elastic stresses are negligible. A thorough analysis of the diffusive growth of these droplets has been given by Lifshitz and Slyozov [1] (who also consider anisotropic effects). Since their paper is a complicated exercise in the asymptotic analysis of coupled, nonlinear dynamical equations, we present only the basic ideas and major results, following the discussion of Lifshitz and Pitaevskii [2]. It is useful to first note that the classical theory of nucleation discussed in Chapter 4 predicts that the critical radius for this solid solution is given by (we consider only the three dimensional case here) :

$$R_c = \frac{2\sigma v'}{\mu' - \mu'_o} \ .$$

(8.1)

For this case μ'_o and v' are the chemical potential and molecular volume of the droplet material, μ' is the chemical potential of the solute in solution and σ is the surface tension for weak solutions.

$$\mu' - \mu'_o = k_B T \ \frac{(\bar{c} - c_{o\infty})}{c_{o\infty}}$$

(8.2)

---

[*] A similar treatment has been given by C. Wagner, Zeit. Electrochem. <u>65</u>, 581 (1961).

where $\bar{c}$ is the mean concentration of the supersaturation solution and $c_{o\infty}$ is the concentration of the saturated solution above a plane sur-face of solute. Thus the critical radius can be written as

$$R_c = \frac{2\sigma v' c_{o\infty}}{k_B T(\bar{c} - c_{o\infty})} \ . \tag{8.3}$$

Alternatively, one could say that the saturation concentration c_{oR} about a spherical droplet of radius R is

$$c_{oR} = c_{o\infty} \left(1 + \frac{2\sigma v'}{k_B TR}\right) \ . \tag{8.4}$$

Returning now to the late stage growth problem, we note that for the very small supersaturation considered, the droplets are far apart as compared with their average size. One can thus ignore the interactions between these droplets. A given droplet then grows by diffusion from the surrounding solution. This local concentration $c(\vec{r})$ in the vicin-ity of a droplet of radius R is given by a solution of the diffusion equation

$$\partial_t c(\vec{r}) = D\nabla^2 c(\vec{r}) \tag{8.5}$$

i.e.

$$\partial_t c(r) = D\frac{1}{r}\frac{\partial^2}{\partial r^2}(rc(r)) \ . \tag{8.6}$$

Furthermore, since the late stage growth occurs only after a nearly equilibrium volume fraction of the minority phase has formed, the con-centration gradients which are responsible for the late stage diffusive growth of a droplet are small. Therefore the diffusion rate is so small during the late stage growth that one can approximate the instantaneous value of the concentration field $c(r)$ about a droplet of size $R(t)$ by the solution of the steady state equation (8.6), with $\partial_t c(r) = 0$. Thus $c(r)$ is given by

$$c(r) = \bar{c} - (\bar{c} - c_{oR})\frac{R}{r} \ . \tag{8.7}$$

Since the concentration is defined in terms of the volume of the material dissolved in unit volume of the solution, the diffusive flux

$$j = D\frac{\partial c(r)}{\partial r}\bigg|_{r=R} = \frac{D}{R}(\bar{c} - c_{oR}) \tag{8.8}$$

at a grain surface is equal to the rate of change of the grain radius, i.e.

$$j = \frac{dR}{dt} = \frac{D}{R}(\bar{c} - c_{oR}) \ . \tag{8.9}$$

Thus we obtain the first of the three basic equations of the LS theory :

$$\frac{dR}{dt} = \frac{D}{R}(\Delta(t) - \frac{\alpha}{R}) \ , \tag{8.10}$$

where $\alpha = (2\sigma v'c_{o\infty})/k_B T$ and $\Delta(t) = \bar{c} - c_{o\infty}$ is the supersaturation of the solution. Note that for every value of the supersaturation $\Delta(t)$, there is a critical radius

$$R_c(t) = \frac{\alpha}{\Delta(t)} \tag{8.11}$$

at which a droplet is in (unstable) equilibrium with the solution. Droplets for which $R \gtrless R_c(t)$ grow or dissolve, respectively. This is the diffusive mechanism by which larger droplets grow at the expense of smaller ones.

A second equation of the LS theory expresses the conservation of the solute molecules, i.e.

$$Q_o = \Delta_o + q_o = \Delta(t) + q(t) \ . \tag{8.12}$$

Here Q_o is the _total_ initial supersaturation including a term q_o which is the volume of material initially in the droplets. The time dependent volume $q(t)$ of the precipitated droplet per unit volume of the solution can be expressed in terms of the droplet size distribution function $\varphi(R,t)$, which is normalized so that

$$N(t) = \int_0^\infty \varphi(R,t)\,dR \tag{8.13}$$

is the number of droplets per unit volume. Thus

$$q(t) = \frac{4\pi}{3} \int R^3 \varphi(R,t) \, dR \ . \tag{8.14}$$

Thus the conservation law (8.12) can be written as

$$1 = \frac{\Delta(t)}{Q_o} + \frac{4\pi}{3Q_o} \int_o^\infty R^3 \varphi(R,t) \, dR \ . \tag{8.15}$$

The third equation of the LS theory is a continuity equation for the droplet distribution function

$$\frac{\partial \varphi}{\partial t} + \frac{\partial}{\partial R}(\varphi v_R) = 0 \ , \tag{8.16}$$

where one considers $v_R = dR/dt$ as the velocity of the droplet in "size space". Note that it is assumed in (8.16) that nucleation has ceased, as otherwise there would be a source term on the right-hand side of (8.16). The three coupled, nonlinear equations (8.10), (8.15) and (8.16) form the starting point of the LS theory. In the LS paper an asymptotic analysis of the solution of these equations is presented for $t \to \infty$ whose results we summarize below. It is worth noting, however, at this stage that the LS theory is in a sense a mean field theory. Direct interactions between droplets are not included, but the effect of the other droplets on the growth of a particular droplet is contained in (8.10) through the supersaturation $\Delta(t)$, which in turn depends on the volume of all the droplets, $q(t)$, and the droplet distribution function $\varphi(R,t)$ through (8.15) and (8.16). For larger initial supersaturations this "mean field" theory breaks down, but a solution of this more complicated problem has not yet been obtained.

8.2 Asymptotic Solution

We summarize the essential features of the analysis of the late stage growth as given by Lifshitz and Slyozov. We introduce the dimensionless variable

$$x(t) = R_c(t)/R_c(0) \tag{8.17}$$

and a "time variable"

$$\tau = 3 \log x(t) \ . \tag{8.18}$$

We know that the asymptotic behavior of $\Delta(t)$ and $R_c(t)$ are given by

$$\Delta(t) \xrightarrow[t\to\infty]{} 0 \ , \tag{8.19}$$

$$R_c(t) \xrightarrow[t\to\infty]{} \infty \ , \tag{8.20}$$

so that $\tau(t)$ varies monotonically from 0 to ∞ in the domain of interest. We next rewrite the equation for droplet growth, (8.10), in terms of the variable

$$u = R/R_c(t) \tag{8.21}$$

(measuring t in (8.10) in units of $R_c^3(0)/D\alpha$). This yields

$$\frac{du^3}{d\tau} = \gamma(u - 1) - u^3 \tag{8.22}$$

where

$$\gamma^{-1} = \gamma^{-1}(\tau) \equiv x^2 \frac{dx}{dt} > 0 \ . \tag{8.23}$$

The key point in the analysis of the asymptotic solution of (8.22) is that $\gamma(\tau)$ approaches a finite number, $\gamma_o = 27/4$, as $\tau \to \infty$ and that it approaches this value from below (see Fig. 8.1b), i.e. :

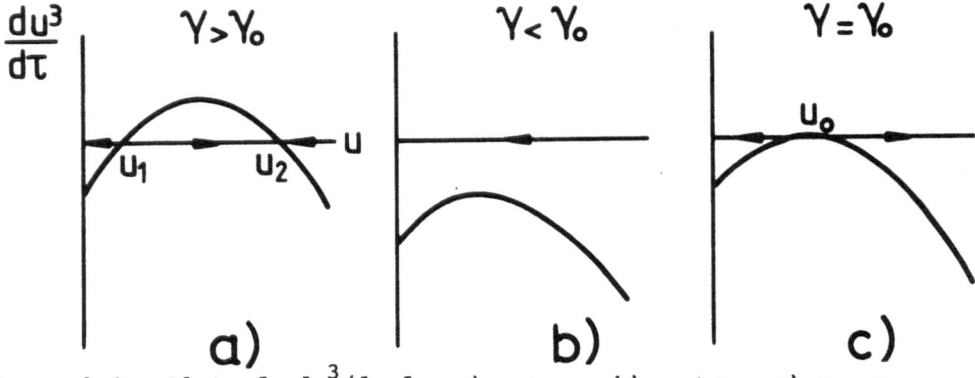

Figure 8.1 : Plot of $du^3/d\tau$ for a) $\gamma > \gamma_o$; b) $\gamma < \gamma_o$; c) $\gamma = \gamma_o$.

$$\gamma(\tau) = \frac{27}{4}\{1 - \frac{3}{4}\frac{1}{\tau^2}[1 + \frac{1}{(\ln\tau)^2}(1+\dots)]\} .$$ (8.24)

(When $\gamma = \gamma_o = 27/4$, $du^3/d\tau = 0$ at $u = u_o = 3/2$.) The other possibilities shown above in Fig. 8.1a,c for which $\gamma > \gamma_o$ or $\gamma(\tau) = \gamma_o$ exactly, can be shown to be inconsistent with the conservation law (8.12) (with $q(t)$ given by (8.14)). When $\tau^2 \gg 1$ the correction term in (8.24) can be ignored and one then obtains the LS asymptotic growth law from (8.23).

$$x^3(t) = (R_c(t)/R_c(0))^3 = 4t/9 .$$ (8.25)

Corrections to this can be worked out and are given by [1]

$$x^3(t) = \frac{4}{9}t[1 + \frac{3}{4(\ln t)^2}(1+\dots)] .$$ (8.26)

One can also show that the droplet distribution function is given by

$$\varphi(R,t) = \frac{A}{R_c^4(t)} R_c^3(0)P(R/R_c(t))$$ (8.27)

where

$$P(u) = \begin{cases} \frac{3^4 \cdot e \ u^2}{2^{5/3}(u+3)^{7/3}(\frac{3}{2}-u)^{11/3}} \exp\{-1/(1-\frac{2}{3}u)\}, & u < 3/2 \\ \\ 0 & , \ u > 3/2 \end{cases}$$ (8.28)

and A is about $(0.9) \times (3Q_o/4\pi R_c^3(0))$. The function $P(u)$ is shown in Fig. 8.2.

The number of grains per unit volume follows from (8.13) and (8.27) as

$$N = \frac{9A}{4t} = \frac{0.5Q_o}{D\alpha t} .$$ (8.29)

One can also show that the mean radius is equal to the critical radius, i.e. $\bar{R}(t) = R_c(t)$. Thus in terms of the original variables one has

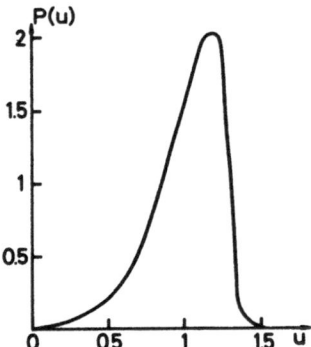

Figure 8.2 : Plot of the function $P(u)$ defined by Equation (8.28).

$$R_c(t) = (\frac{4\alpha Dt}{9})^{1/3} \ . \tag{8.30}$$

The vanishing of the supersaturation is given by

$$\Delta(t) = (\frac{9\alpha^2}{4Dt})^{1/3} \ . \tag{8.31}$$

REFERENCES   -   Chapter 8.

[1] I.M. Lifshitz and V.V. Slyozov, J. Phys. Chem. Solids 19, 35
(1961).

[2] I.M. Lifshitz and L.P. Pitaevskii, "Physical Kinetics", Landau
and Lifshitz Course of Theoretical Physics, vol. 10, Pergamon
Press (London) (1981).

In this chapter we discuss the dynamics of interfaces for a system whose order parameter is nonconserved, based on the so-called kinetic drumhead model. This is a dynamical version of the static drumhead model discussed in Chapter 5. The deterministic version of this dynamical model was first derived by Allen and Cahn [1] and used to discuss domain coarsening in solid systems for which the order parameter is not conserved. We will review their work in Section 9.1. We present in 9.2 a derivation of the Langevin equation for the drumhead model due to Kawasaki and Ohta [2] in the limit of an infinitely deep potential well for the order parameter. Their derivation provides in principle the possibility of a systematic improvement of the kinetic drumhead model. An equivalent derivation of the Langevin equation for the drumhead model has been given by Bausch et al. [3], who used it to discuss the critical dynamics of model A in a $1 + \epsilon$ expansion. In general this model is a useful starting point for problems in which the intrinsic structure of the interface itself is not important.

9.1 Allen-Cahn Deterministic Equation

Before discussing a systematic derivation of the Langevin equation for the kinetic drumhead model, we present the Allen-Cahn derivation of the deterministic version of this model. Their derivation is very transparent, although it is not as systematic as the derivation given in 9.2.

Allen and Cahn consider the Langevin equation (3.23) for model A, but neglect the noise term ξ. They assume a general Ginzburg-Landau form (3.24) for the free energy functional, but without restricting the potential to a fourth-order polynomial. Thus

$$F = \int d^d\vec{x} \{ \tfrac{1}{2} c |\nabla c|^2 + V(c) \} \tag{9.1}$$

where $V(c) = V(-c)$. In mean field theory the location of the minima of $V(c)$, $\pm c_o$, are the equilibrium values of the order parameter as discussed in Chapter 5. Also recall from Chapter 5 that the mean field theory for planar interfaces is given by the solution of

$$\frac{\delta F}{\delta c} = - C \frac{\partial^2 c}{\partial \bar{z}^2} + \frac{\partial V}{\partial c} = 0 \qquad (9.2)$$

where $\bar{z}$ is the distance along the unit normal $\vec{n}$ to the interface (see Eqs (5.3), (5.17) and Fig. 5.1).

With these elementary remarks we now consider the deterministic form of (3.23), i.e.

$$\frac{\partial c}{\partial t} = - \Gamma_o (\frac{\partial V}{\partial c} - C \nabla^2 c) . \qquad (9.3)$$

Now suppose that the "late stage" non equilibrium situation described by (9.3) consists of regions in which the order parameter c is close to one of its two (mean field) equilibrium values $\pm c_o$. A region of "plus" phase c_o is separated from a region of "negative" phase $-c_o$ by an interface. It is then natural to describe the dynamical evolution of this system in terms of the random motion of the interfaces, whose equation of motion can be obtained (under certain restrictions) from (9.3). To obtain the equation for the normal component v of an interface velocity, we consider the case in which the principal radii of curvature of the interface are large compared to the interface thickness (which in equilibrium is given by the correlation length ξ). Assume further that the interfaces are gently curved, so that one can treat these interfaces as locally planar. Given these restrictions, we can derive an equation for v by using some simple identities and equation (9.2).

Let us first introduce the natural coordinate system for an interface and, for example, write

$$\nabla c = \vec{n} \frac{\partial c}{\partial \bar{z}} \qquad (9.4)$$

where $\partial c / \partial \bar{z}$ is the rate of change of c in the direction of the unit normal $\vec{n}$. One can also show that $\nabla^2 c$ (which appears in (9.3)) can be written as [1,5]

$$\nabla^2 c = \nabla \cdot \nabla c = \frac{\partial^2 c}{\partial \bar{z}^2} + \frac{\partial c}{\partial \bar{z}} (\nabla \cdot \hat{n}) \qquad (9.5)$$

$$\nabla \cdot \vec{n} = -(K_1 + K_2) \equiv -K \qquad (9.6)$$

where K_1 and K_2 are the two principal curvatures of the surface and K is the mean curvature. Thus we can rewrite the deterministic Langevin equation (9.3) in terms of the interface variables as

$$\frac{\partial c}{\partial t} = -\Gamma_o \{ \frac{\partial V}{\partial c} - C[\frac{\partial^2 c}{\partial \bar{z}^2} - (K_1 + K_2)\frac{\partial c}{\partial \bar{z}}] \} . \qquad (9.7)$$

(Note that we can always divide our system into surfaces of constant c and introduce the above coordinate system.) We now use (9.7) to obtain an equation for the interface velocity. Consider the motion of a gently curved interface which at some particular time has the profile $c_p(\bar{z})$ at all normal sections. Then in this case (9.2) is valid, so (9.7) reduces to

$$(\frac{\partial c}{\partial t})_{\bar{z}} = - M(K_1 + K_2) (\frac{\partial c}{\partial \bar{z}})_t \qquad (9.8)$$

where

$$M \approx C \Gamma_o . \qquad (9.9)$$

Thus the velocity $(\partial \bar{z}/\partial t)$ of a constant c surface in the interface region is

$$(\frac{\partial \bar{z}}{\partial t}) = -(\frac{\partial c}{\partial t})_{\bar{z}} / (\frac{\partial c}{\partial \bar{z}})_t . \qquad (9.10)$$

Finally, if the principal radii of curvature are much larger than the interface thickness, the curvatures of the constant-c surfaces are independent of the value of the coordinate $\bar{z}$. Thus all surfaces of constant-c at a point in the interface will move with the same velocity v, which from (9.8) and (9.10) is

$$v = M(K_1 + K_2) . \qquad (9.11)$$

This is the Allen-Cahn result. Note also the initial profile $c_p(\bar{z})$ is preserved in the moving interface.

9.2 Langevin Equations

We start with the Ginzburg-Landau equation of motion for the non-conserved case (model A) discussed in Chapter 3, i.e.

$$\partial_t c(\vec{x},t) = -\Gamma_o \frac{\delta F}{\delta c(\vec{x},t)} + \zeta(\vec{x},t) \tag{3.23}$$

where

$$<\zeta(\vec{x},t)> = 0 \text{ and } <\zeta(\vec{x},t)\zeta(\vec{x}',t')> = 2\Gamma_o \delta(\vec{x}-\vec{x}')\delta(t-t') \ . \tag{3.25}$$

Our goal is to derive an equation of motion for the displacement of the interface $f(\vec{r},t)$ from the Ginzburg-Landau model, in the low temperature limit $\tau, g \to \infty$, with the equilibrium mean field value of the order parameter $m_o = (6\tau/g)^{1/2}$, finite (see Section 5.1). One way to proceed is via a path integral method, involving a path probability distribution functional for the order parameter. This has been explicitly carried out by Kawasaki and Ohta, using the formalism of Bausch, Janssen and Wagner [4]. This has the advantage of allowing memory effects, which permits a systematic way to develop correction terms for finite τ and g. It is, however, somewhat more complicated to discuss and we therefore consider a second approach in which memory effects are neglected from the onset. Thus we assume a Langevin equation for the collective coordinate $f(\vec{r},t)$,

$$\partial_t f(\vec{r},t) = v_f(\{f\};\vec{r}) + \theta_f(\vec{r},t) \tag{9.12}$$

with a fluctuation-dissipation relation

$$<\theta_f(\vec{r},t)\theta_f(\vec{r}',t')> = Q\delta(\vec{r}-\vec{r}')\delta(t-t') \ . \tag{9.13}$$

We wish to determine v_f, θ_f and Q from the original model A equations (3.23) and (3.25). As in our treatment of the static drumhead model, we consider a solution of the form

$$c(\vec{x},t) = m_{z_o} (z-f(\vec{r},t)) + \eta(z-f(\vec{r},t),\vec{r},t) \tag{9.14}$$

i.e.

$$c(\vec{x},t) = \chi(z-f(\vec{r},t),\vec{r},t) \ . \tag{9.15}$$

A systematic expansion of model A can then be obtained by taking η to be $0(\tau^{-1/2})$ (Bausch et al. [3]). It should be noted that the fact that the only time dependence of the first term of (9.14) arises from $f(\vec{r},t)$ is the assertion that in the low temperature limit $\tau \sim g \to \infty$, the only important dynamical variable is f.

To obtain an equation of motion for f in this limit, we substitute (9.15) into the Ginzburg-Landau equation (3.23) and obtain

$$\partial_t\chi = \Gamma_0[\ a^2\partial_z^2\chi + \tau\chi - \frac{g}{6}\chi^3] + \partial_z\chi(\dot{f}-\Gamma_0\partial_{\vec{r}}^2 f) - 2\Gamma_0\partial_{\vec{r}}f\cdot\partial_{\vec{r}}\partial_z\chi + \theta \tag{9.16}$$

with

$$\theta(\vec{r},z,t) \equiv \zeta(\vec{r},z+f,t) \ . \tag{9.17}$$

The term $(\partial_z\chi)\dot{f}$ on the right-hand side of (9.16) arises from the fact that the time derivative of (9.15) yields two terms, $\dot{\chi} = \partial_t\chi - \dot{f}\partial_z\chi$, where $\partial_t\chi$ describes the time variation of χ in the coordinate system moving with the interface. We now consider the limit $\tau \sim g \to \infty$ with $m_0 = (6\tau/g)^{1/2}$ fixed. If one divides (9.16) by τ and considers the limit of large τ, it is clear that χ will approach its steady state value in a short time, of order τ^{-1}. Furthermore, the dominant term on the right-hand side of this equation will be the first term (in brackets). Thus a first approximation to the steady-state solution $\partial_t\chi = 0$ in this limit is :

$$\chi(\vec{x},t) = \chi_s(z) = m_{z_0}(z/a) \tag{9.18}$$

where we have used the mean field equation (5.19). (It should be noted that $\chi_s(z)$ depends on $\vec{r}$ through the factor a defined in (5.17) which appears in (9.18).) We can obtain the desired equation of motion for f by considering the next correction term to χ_s, namely

$$\chi(\vec{x},t) = m_{z_0}(z/a) + m(\vec{x},t) \ . \tag{9.19}$$

Substituting (9.19) into (9.16) with $\partial_t\chi = 0$ gives

$$\Gamma_0[a^2\partial_z^2 + \tau - \frac{g}{2}m_{z_0}^2(z/a)]m - \Gamma_0\frac{g}{6}[3m_{z_0}(z/a)m^2 + m^3]$$

$$+ [\partial_z(m_{z_0}(z/a) + m)](\dot{f} - \Gamma_0\partial_r^2 f)$$

$$- 2\Gamma_0\partial_{\vec{r}}f\cdot\partial_{\vec{r}}\partial_z[m_{z_0}(z/a) + m] + \theta = 0 \ . \tag{9.20}$$

Thus considering the fluctuations m as small compared to $m_{z_0}(z/a)$ we obtain the approximate equation

$$\Gamma_0\Omega(z/a)m = u(z) \tag{9.21}$$

where $\Omega(z/a)$ is the operator $\mathbf{M}_0$ defined in (5.54), with z replaced by z/a, i.e.

$$\Omega(z/a) = -a^2\partial_z^2 - \tau + \frac{g}{2}\chi_s(z) \tag{9.22}$$

and the quantity $u(z)$ is

$$u(z) = [\dot{f} - \Gamma_0\partial_r^2 f - 2\Gamma_0\partial_{\vec{r}}f\cdot\partial_{\vec{r}}]\partial_z\chi_s(z) + \theta \ . \tag{9.23}$$

Next, we multiply (9.21) by $\partial_z\chi_s(z)$ and integrate over z to obtain for nonzero m a condition on u. Namely, with $(\varphi(z),\psi(z)) \equiv \int_{-\infty}^{+\infty}dz\varphi(z)\psi(z)$

$$\Gamma_0(\partial_z\chi_s,\Omega(z/a)m) = (\partial_z\chi_s,u(z)) \ . \tag{9.24}$$

But by its definition (9.22), $\Omega(z/a)$ is clearly self-adjoint, so that

$$\Gamma_0(\Omega(z/a)\partial_z\chi_s,m) = (\partial_z\chi_s,u(z)) \ . \tag{9.25}$$

But from Eqs (5.83) and (5.84) $\Omega\partial_z\chi_s = 0$ and so we obtain the explicit condition

$(\partial_z \chi_s, u) = 0$ i.e. :

$$\partial_t f (\partial_z \chi_s, \partial_z \chi_s) - \Gamma_o (\partial_z \chi_s, [\partial_{\vec{r}}^2 f + 2\Gamma_o \partial_{\vec{r}} f \cdot \partial_{\vec{r}}] \partial_z \chi_s) + (\partial_z \chi_s, \theta) = 0 \; . \quad (9.26)$$

Thus using the definition for the surface tension (5.31) we have

$$\frac{\sigma}{aT} [\partial_t f - v_f - \theta_f] = 0 \qquad\qquad (9.27)$$

which yields the Langevin equation (9.12), with

$$\theta_f = -\frac{aT}{\sigma} (\partial_z \chi_s, \theta) \qquad\qquad (9.28)$$

and, with the notation $\|\varphi\|^2 \equiv \int_{-\infty}^{+\infty} \varphi^2 (z) \, dz$,

$$v_f = \Gamma_o [\partial_{\vec{r}}^2 f + \partial_{\vec{r}} f \cdot \partial_{\vec{r}} \ell n \|\partial_z \chi_s\|^2] \; . \qquad\qquad (9.29)$$

The expression for the velocity can be rewritten, using

$$\|\partial_z \chi_s\|^2 = \frac{\sigma}{aT} \qquad\qquad (9.30)$$

as

$$v_f = \Gamma_o [\partial_{\vec{r}}^2 + \frac{1}{a^2} \partial_{\vec{r}} f \cdot \partial_{\vec{r}} \partial_{\vec{r}} f \cdot \partial_{\vec{r}} f] \; . \qquad\qquad (9.31)$$

The quantity Q in (9.13) can be obtained using (9.28), (9.17) and (3.25) as

$$<\theta_f (\vec{r}, t) \theta_f (\vec{r}', t')> = (\frac{aT}{\sigma})^2 \int_{-\infty}^{\infty} \int_{-\infty}^{\infty} dz \, dz' \, \partial_z \chi_s (z) \partial_z \chi_s (z') \cdot$$

$$<\theta (\vec{x}, t) \theta (\vec{x}', t')>$$

$$= 2\Gamma_o (\frac{aT}{\sigma})^2 \delta(\vec{r} - \vec{r}') \delta(t - t') \int_{-\infty}^{\infty} \int_{-\infty}^{\infty} dz \, dz'$$

$$\delta(z - z') \partial_z \chi_s (z) \partial_z \chi_s (z')$$

i.e. $<\theta_f (\vec{r}, t) \theta_f (\vec{r}', t')> = 2\Gamma_o \frac{aT}{\sigma} \delta(t - t') \delta(\vec{r} - \vec{r}')$ \qquad (9.32)

and thus :

$$Q = 2\Gamma_o aT/\sigma \ . \tag{9.33}$$

There is in fact a more natural way to write the Langevin equation (9.12), namely in terms of the component of the velocity normal to the interface. That is, we have seen in Sections 5.1 and 9.1 that the natural variable is $\bar{z} = z/a$. Thus $\delta\bar{z} = \delta z/a = \delta f/a$, so (9.12) becomes

$$a^{-1} \frac{\partial f}{\partial t} = a^{-1} v_f + \zeta_f \tag{9.34}$$

where $\zeta_f = \theta_f/a$ satisfies

$$<\zeta_f(\vec{r},t)\,\zeta_f(\vec{r}',t')> = \frac{2\Gamma_o T}{\sigma a}\delta(\vec{r}-\vec{r}')\delta(t-t') \ . \tag{9.35}$$

If we introduce the invariant mean curvature of the interface as

$$K = -\vec{\nabla}\vec{n} = \partial_{\vec{r}}\cdot(\frac{\partial_{\vec{r}}f}{a}) \tag{9.36}$$

then it is straightforward to show that

$$K = -\frac{1}{\sigma}\frac{\delta F_{dh}\{f\}}{\delta f(\vec{r})} \tag{9.37}$$

where $F_{dh}\{f\}$ is the free energy of the static drumhead model, (5.50). It also follows from (9.31) and (9.36) that

$$K = a^{-1} v_f \ . \tag{9.38}$$

Thus we can write the Langevin equation as

$$a^{-1} \frac{\partial f}{\partial t} = \Gamma_o K + \zeta_f \ . \tag{9.39}$$

As noted above, the left-hand side, $a^{-1}\partial f/\partial t$, is the velocity v perpendicular to the interface. Thus (9.39) is the statement that the local mean curvature K provides the "driving force" for the "perpendicular" velocity of the interface. This equation is slightly subtle, since one might expect the driving force to be given by the change in the free energy, $\delta F/\delta f$, associated with the change in area of the surface at $\vec{x}$. However, as can be seen from (9.37) $1/\sigma \; \delta F_{dh}/\delta f$ is the driving force, so that the surface tension does not enter (9.34). Its only role is in the fluctuation-dissipation expression for the noise, ζ_f, in (9.35).

We conclude this section by briefly mentioning other properties related to the kinetic drumhead model. First, Kawasaki and Ohta have shown that the corresponding Fokker-Planck equation for the time dependent probability distribution functional $P(\{f\},t)$ for the collective coordinate f is

$$\partial_t P(\{f\},t) = -\int d\vec{r} \; \delta J/\delta f(\vec{r}) \tag{9.40}$$

where

$$J = \frac{\Gamma_o T}{\sigma} a \; [\frac{\delta}{\delta f(\vec{r})} + \frac{\delta F_{dh}\{f\}}{\delta f(\vec{r})}] P(\{f\},t) \; . \tag{9.41}$$

The equilibrium distribution functional P_e has the form

$$P_e(\{f\}) = N e^{-F_{dh}\{f\}} \tag{9.42}$$

where N is an appropriate normalization constant.

Second, both the Langevin equation (9.39) and the Fokker-Planck equation (9.40) have been shown by Kawasaki and Ohta to be invariant under the Euclidean group of transformations. That this should be the case for the kinetic drumhead model follows from a reasoning similar to the one presented in Chapter 5 for the static drumhead model. Namely, the original free energy (5.1) and the Langevin and Fokker-Planck equations for model A (3.23) and (3.26) are invariant under the Euclidean group of transformations. However, the interfaces destroy these invariances in the original model. By finding suitable dynamical equations of motion for the variable $f(\vec{r},t)$, it is possible to restore this invariance, just as the drumhead free energy $F_{dh}\{f\}$ is invariant under the Euclidean group (Chapter 5). We refer the reader to Reference [2] for the details of the proof for the kinetic model.

REFERENCES - Chapter 9.

[1] S.M. Allen and J.W. Cahn, Acta. Metall. 27, 1085 (1979).

[2] K. Kawasaki and T. Ohta, Prog. Theor. Phys. 67, 147 (1982).

[3] R. Bausch, V. Dohm, H.K. Janssen and R.K.P. Zia, Phys. Rev. Lett. 47, 1837 (1981).

[4] R. Bausch, H.K. Janssen and H. Wagner, Zeit. Physik B24, 113 (1976).

[5] G. Arfken, in "Mathematical Methods for Physicists", p. 65-69, Academic Press, New York, (1968).

[6] C. Weatherburn, in "Differential Geometry of Three Dimensions", p. 225, Cambridge University Press, Oxford, (1927).

GENERAL REFERENCES

[G1] K. Kawasaki and T. Ohta, in "Conference on Nonlinear Fluid Behaviour", Boulder, Colorado June 7-11 (1982), to be published. This paper discusses interface dynamics for systems with a conserved order parameter (such as for a binary alloy). It also includes a treatment of the effects of hydrodynamics, such as arises in fluids.

[G2] S.K. Chan, J. Chem. Phys. 67, 5755 (1977). This paper is an interesting treatment of interface motion in which the shape-invariant profile of the interface (Section 9.1) is explicitly discussed.

CHAPTER 10. DYNAMICAL SCALING

One of the most interesting discoveries in recent years has been that the structure factor $S(k,t)$ discussed in earlier chapters satisfies (to a reasonable first approximation) a scaling law of the form

$$S(k,t) = \bar{R}^d(t) F(k\bar{R}(t)), \quad t > t_o \tag{10.1}$$

where $\bar{R}(t)$ is a characteristic length (which is system dependent) and t_o is some initial "transient" time. This behavior has been observed in relaxational type systems for which the order parameter is nonconserved (e.g. model A) or conserved (e.g. model B), as well as in systems in which hydrodynamics is important, such as binary fluids. A discussion of a variety of such systems for which (10.1) holds, as well as a review of various phenomenological theories which lead to (10.1) is given by Gunton, San Miguel and Sahni [1]. In this chapter we summarize what appears to be a promising first principles theory for calculating $S(k,t)$, which is based on the kinetic drumhead model of the preceding chapter. Formulating the problem in terms of the motion of random interfaces seems a very useful approach for understanding the "late stage" dynamics (including (10.1)) of quenched systems. Our discussion here deals with recent work by Ohta, Jasnow and Kawasaki [2] on the dynamics of random interfaces for systems quenched below an order-disorder transition temperature.

Ohta, Jasnow and Kawasaki obtain an approximate solution of the Allen-Cahn equation in d-dimensions for the normal component of an interface velocity, given by (9.11), i.e.

$$v(\vec{r}(a,t)) = MK(\vec{r}(a,t)) \tag{10.2}$$

where $\vec{r}(a,t)$ denotes the position of a point on the interface at time t. The result of their approximate solution is a prediction for the average area density $A(t)$ of interfaces and for the dynamic structure function $S(k,t)$. In particular the authors derive equation (10.1), with an explicit form for the scaling function $F(k\bar{R}(t))$ which is in reasonable agreement with Monte Carlo studies of the $d = 2$ and $d = 3$ kinetic antiferromagnetic Ising model. Their first step is to rewrite

(10.2) in a more convenient form for analysis. To do this they introduce a surface $u(\vec{r},t) = z'$ where $\vec{r}$ is a d-dimensional vector. The
interfaces for the order-disorder problem described by (10.2) are taken
to be given by the intersection of this surface (assumed smooth) with
the plane $z' = 0$. Introducing this surface is simply a useful mathematical way in which to describe the physically relevant interfaces
given by $u(\vec{r},t) = 0$. The normal unit vector $n(\vec{r},t)$ on the inter-
faces is thus given by $\vec{n}(r,t) = \nabla u/|\nabla u|$. Since $du(\vec{r},t)/dt = 0$ in
the reference frame moving with the interface, one obtains for the nor-
mal component of the interface velocity

$$v(\vec{r},t) = -\frac{1}{|\nabla u|} \frac{\partial u(\vec{r},t)}{\partial t} . \tag{10.3}$$

Since the curvature K of the interface is given by $K = -\nabla \cdot \vec{n}$, (10.2)
can be written as

$$\frac{1}{|\nabla u|} \frac{\partial u(\vec{r},t)}{\partial t} = M\nabla \cdot \left(\frac{\nabla u}{|\nabla u|}\right) . \tag{10.4}$$

It is straightforward to evaluate the divergence on the right-hand side
of (10.4) to obtain

$$\frac{\partial u(\vec{r},t)}{\partial t} = M\{\nabla^2 - \sum_{\alpha,\beta} n^\alpha(\vec{r},t) n^\beta(\vec{r},t) \nabla^\alpha \nabla^\beta\} u(\vec{r},t) \tag{10.5}$$

where $\alpha, \beta = 1,\ldots,d$ denote the Cartesian components of $\vec{n}$ and ∇. A
configuration of interfaces then follows formally from the solution
$u(\vec{r},t)$ of (10.5) for a given initial surface (and corresponding initial interface configuration) $u_o(\vec{r}) = u(\vec{r},0)$, by setting $u(\vec{r},t) = 0$.
One does not have a unique initial $u_o(\vec{r})$ but rather some probability
distribution of initial configurations. (Recall that the exact equation
of motion for the drumhead model is stochastic.) Ohta et al. make the
simplest possible assumption, namely that this initial probability distribution $P(\{u_o\})$ is Gaussian. They further assume that the two non-
zero correlation functions are

$$\langle u_o(\vec{r}) \rangle = U \tag{10.6}$$

and

$$\langle \delta u_o(\vec{r}) \delta u_o(\vec{r'}) \rangle = B\delta(\vec{r}-\vec{r'}) \tag{10.7}$$

where < > denotes an average with respect to P and $\delta u = u - <u>$. Although Ohta et al. considered the case $U \neq 0$, we will restrict our-selves to $U = 0$. As noted by them, physically observable quantities should be independent of the choice of contour surfaces $u(\vec{r},t) = $ con-stant, which requires $U = 0$ in this case.

The quantities $A(t)$ and $S(k,t)$ can be described formally in terms of the field $u(\vec{r},t)$. Thus, for example, the average area den-sity of the interfaces is given by

$$A(t) = \int da <\delta(\vec{r}-\vec{r}(a))> \tag{10.8}$$

where the integral over a is taken over all interfaces. Using the generalization of the identity for a delta function, i.e.

$$\delta(f(x)) = \sum_{x_i} \delta(x-x_i)/|(\frac{\partial f}{\partial x})_{x_i}| , \tag{10.9}$$

where x_i denotes the zeroes of $f(x)$, one can write (10.8) in terms of the field $u(\vec{r},t)$ as

$$A(t) = <|\nabla u|\delta(u(\vec{r},t))> . \tag{10.10}$$

One can also shwo that [3] :

$$S(k,t) = \frac{(\Delta m_o)^2}{k^2} \int d^d(\vec{r}_1-\vec{r}_2) \sum_{\alpha} <[u_\alpha(\vec{r}_1,t)][u_\alpha(\vec{r}_2,t)]\delta(u(\vec{r}_1,t))$$
$$\delta(u(\vec{r}_2,t))>e^{i\vec{k}\cdot(\vec{r}_1-\vec{r}_2)} \tag{10.11}$$

where u_α denotes the α-th component of the gradient of u and Δm_o is the "gap" of the order-parameter profile at an interface. We derive (10.11) later.

Given this stochastic description for the interface motion, Ohta et al. then linearize the equation of motion (10.5). The justification for this can only be in the a posteriori comparison with experiment, since there is no obvious reason that this should be a valid approxima-tion. (One could, however, view this as the first approximation to a perturbation theory treatment of (10.5).) The linearization follows by making the approximation

$$n^\alpha(\vec{r},t)n^\beta(\vec{r},t) \cong <n^\alpha(\vec{r},t)n^\beta(\vec{r},t)> \qquad (10.12)$$

$$= \frac{1}{d}\delta_{\alpha\beta} . \qquad (10.13)$$

(In going from (10.12) to (10.13) we have assumed that the system is isotropic.) One could then treat the difference $n^\alpha n^\beta - <n^\alpha n^\beta>$ by a perturbation expansion. Given (10.13), one obtains from (10.5) the linear equation

$$\frac{\partial u(\vec{r},t)}{\partial t} = L\nabla^2 u(\vec{r},t) \qquad (10.14)$$

with $L = M(d-1)/d$. This diffusion-like equation can then be solved by standard means [*]. The solution for $d = 3$ is

$$u(\vec{r},t) = \frac{1}{\pi^{3/2}\bar{R}^3} \int_{-\infty}^{\infty}\int_{-\infty}^{\infty}\int_{-\infty}^{\infty} dx'dy'dz' u_0(\vec{r}')e^{-(\vec{r}-\vec{r}')^2/\bar{R}^2} \qquad (10.15)$$

where $\bar{R} = (4Lt)^{1/2}$ is the average domain size predicted in the Allen-Cahn theory and $\vec{r} = (x,y,z)$. To evaluate $A(t)$ from (10.10) one also needs the gradient of $u(\vec{r},t)$, which follows from (10.15) as

$$\nabla u(\vec{r},t) = \frac{2}{\pi^{3/2}\bar{R}^5} \int_{-\infty}^{\infty}\int_{-\infty}^{\infty}\int_{-\infty}^{\infty} dx'dy'dz' [\vec{r}'-\vec{r}]u_0(\vec{r}') \cdot$$

$$\cdot e^{-(\vec{r}-\vec{r}')^2/\bar{R}^2} . \qquad (10.16)$$

It is sometimes convenient to express the integrals in (10.15) and (10.16) in scaled form by writing

$$x_1 = (x' - x)/\bar{R}$$
$$x_2 = (y' - y)/\bar{R} \qquad (10.17)$$
$$x_3 = (z' - z)/\bar{R} .$$

Thus :

---

[*] cf. for example H. Margenau and G.M. Murphy, "Mathematics of Physics and Chemistry", ed. Van Nostrand, New York (1951).

$$u(\vec{r},t) = \frac{1}{\pi^{3/2}} \int_{-\infty}^{\infty}\int_{-\infty}^{\infty}\int_{-\infty}^{\infty} dx_1 dx_2 dx_3 u_o(\bar{R}\vec{x} + \vec{r}) e^{-\vec{x}^2} \tag{10.18}$$

$$\nabla u(\vec{r},t) = \frac{2}{\pi^{3/2}\bar{R}} \int_{-\infty}^{\infty}\int_{-\infty}^{\infty}\int_{-\infty}^{\infty} dx_1 dx_2 dx_3 u_o(\bar{R}\vec{x} + \vec{r}) \cdot e^{-\vec{x}^2}$$

$$[x_1 \hat{\imath} + x_2 \hat{\jmath} + x_3 \hat{k}] \tag{10.19}$$

with $\vec{x} = (x_1, x_2, x_3)$.

One quantity of mathematical interest is the time correlation function of the field $u(\vec{r},t)$:

$$G(\vec{r},t) = \langle u(\vec{r},t) u(\vec{0},t) \rangle . \tag{10.20}$$

It follows from (10.15) and (10.20), together with (10.7) (with $U = 0$) that

$$G(\vec{r},t) = \frac{B}{\pi^3 \bar{R}^6} e^{-r^2/2\bar{R}^2} \int_{-\infty}^{\infty} dx' e^{-2(x'-x/2)^2/\bar{R}^2} .$$

$$\cdot \int_{-\infty}^{\infty} dy' e^{-2(y'-y/2)^2/\bar{R}^2} \int_{-\infty}^{\infty} dz' e^{-2(z'-z/2)^2/\bar{R}^2} , \tag{10.21}$$

i.e.

$$G(\vec{r},t) = \frac{B}{(2\pi)^{3/2}\bar{R}^3} e^{-r^2/2\bar{R}^2} . \tag{10.22}$$

To obtain $A(t)$ from (10.10), we first rewrite this equation as

$$A(t) = \langle \frac{(\nabla u)^2}{|\nabla u|} \delta(u(\vec{r},t)) \rangle \tag{10.23}$$

and then use the identity

$$\frac{1}{|\nabla u|} = 4\pi \int \frac{d^3\vec{q}}{(2\pi)^3} \frac{e^{i\vec{q}\cdot\nabla u}}{q^2} \tag{10.24}$$

to obtain

$$A(t) = 4\pi \int \frac{d^3\vec{q}}{(2\pi)^3} \frac{1}{q^2} \langle (\nabla u)^2 \delta(u(\vec{r},t)) e^{i\vec{q}\cdot\nabla u} \rangle . \tag{10.25}$$

To proceed further we introduce the generating functional

$$Z = <\delta(u(\vec{r},t))e^{i\vec{q}\cdot\nabla u}> . \tag{10.26}$$

It follows upon differentiation that the quantity which enters (10.25) is given by

$$-\nabla^2_{\vec{q}}Z = <(\nabla u)^2\delta(u(\vec{r},t))e^{i\vec{q}\cdot\nabla u}> . \tag{10.27}$$

The generating functional Z can be evaluated by the method of cumulants, since the distribution function is Gaussian. One way to handle the δ-function is to introduce the representation

$$\delta(u(\vec{r},t)) = \lim_{n\to\infty} \frac{n}{\sqrt{\pi}} \int_{-\infty}^{\infty} e^{-n^2u^2}du . \tag{10.28}$$

Making use of the property of a Gaussian distribution [*], namely

$$<e^{i\vec{q}\cdot\nabla u}> = e^{<(i\vec{q}\cdot\nabla u)>+1/2<(i\vec{q}\cdot\nabla u)^2>_c} \tag{10.29}$$

and the averages (which follow from (10.18) and (10.19)) for $d = 3$

$$<u^2> = \frac{1}{(2\pi)^{3/2}} \frac{B}{R^3} \tag{10.30}$$

$$<u^2_\alpha> = \frac{1}{(2\pi)^{3/2}} \frac{B}{R^5} , \quad \alpha = x,y,z \tag{10.31}$$

(where $u_x = \partial u/\partial x,\ldots$) one can evaluate (10.26) and (10.27). Upon performing the remaining integral in (10.25), one obtains for the average area density in $d = 3$

$$A(t) = \frac{2\sqrt{3}}{\pi} \frac{1}{\bar{R}(t)} . \tag{10.32}$$

The result that $A(t) \propto \bar{R}^{-1}(t)$ was first derived by Allen and Cahn in their 1979 paper [4], but without determining the constant of propor-

_____

[*] see for example R. Fox, Physics Reports __48__, 18 (1978).

tionality.

The calculation of the structure function $S(k,t)$ is more complicated, as we discuss before. We first derive (10.11) from the definition

$$S(\vec{k},t) = \int d^d(\vec{r}_1-\vec{r}_2) e^{i\vec{k}\cdot(\vec{r}_1-\vec{r}_2)} <c(\vec{r}_1,t)c(\vec{r}_2,t)> \qquad (10.33)$$

where $c(\vec{r},t)$ is the order parameter. We can rewrite this as

$$S(\vec{k},t) = \int d^d(\vec{r}_1-\vec{r}_2) \frac{e^{i\vec{k}\cdot(\vec{r}_1-\vec{r}_2)}}{k^2} <\nabla_1 c(\vec{r}_1,t)\cdot\nabla_2 c(\vec{r}_2,t)> \qquad (10.34)$$

since (10.34) yields (10.33) upon integration by parts. Next, we follow the procedure of Chapter 9 and decompose $c(\vec{r},t)$ into its mean field value m_o and a small correction term m,

$$c(\vec{r},t) = m_o(u(\vec{r},t)) + m(\vec{r},t) \qquad (10.35)$$

where the position of the interface is given by $u(\vec{r},t) = 0$. Then in the same approximation as used in Chapter 9, we can write

$$\nabla c(\vec{r},t) \simeq \nabla m_o(u(\vec{r},t)) \cong \nabla u(\vec{r},t) \Delta m_o \delta(u(\vec{r},t)) \quad . \qquad (10.36)$$

The result (10.36) holds for the limiting case of the infinitesimally thin interface, with Δm_o denoting the discontinuity in the equilibrium value of the order parameter across the interface. Inserting (10.36) into (10.34) then leads to (10.11).

To evaluate [*] $S(\vec{k},t)$ from (10.11) we need to calculate

$$S_{\alpha\alpha}(\vec{r}_1-\vec{r}_2,t) \equiv <\nabla_1^\alpha u(\vec{r}_1,t) \nabla_2^\alpha u(\vec{r}_2,t) \delta u(\vec{r}_1,t) \delta u(\vec{r}_2,t)> \qquad (10.37)$$

where α denotes the cartesian coordinates in d-dimensions. We first note that the d-dimensional generalization of (10.22) is

---

[*] We are indebted to Dr. Martin Grant for the following derivation.

$$\langle u(\vec{r}_1,t)u(\vec{r}_2,t)\rangle = \sigma^2 \bar{R}^2 e^{-R^2/2} \tag{10.38}$$

where $\vec{R} \equiv (\vec{r}_1-\vec{r}_2)/\bar{R}$ and

$$\sigma^2 \equiv \frac{B}{(2\pi)^{d/2}\bar{R}^{d+2}} \, . \tag{10.39}$$

It follows by taking appropriate derivatives of (10.38) that

$$\langle \nabla_1^\alpha u(\vec{r}_1,t)u(\vec{r}_2,t)\rangle = -R^\alpha \sigma^2 \bar{R} e^{-R^2/2} \tag{10.40}$$

and

$$\langle \nabla_1^\alpha u(\vec{r}_1,t)\nabla_2^\beta u(\vec{r}_2,t)\rangle = (\delta^{\alpha\beta}-R^\alpha R^\beta)\sigma^2 e^{-R^2/2} \tag{10.41}$$

in an obvious notation. We can also write S_{xx} in (10.37) as

$$S_{xx} = \frac{1}{\bar{R}^2}\langle \Delta_1 \Delta_2 \delta(u_1)\delta(u_2)\rangle \tag{10.42}$$

where

$$\Delta_i \equiv \frac{1}{\sigma} \nabla_i^x u(\vec{r}_i,t) \, , \quad i = 1,2 \tag{10.43}$$

$$u_i \equiv \frac{1}{\sigma\bar{R}} u(\vec{r}_i,t) \, , \quad i = 1,2 \, . \tag{10.44}$$

The correlations of these new variables follow from their definitions and (10.38)-(10.41) as

$$\langle \Delta_i^2 \rangle = \langle u_i^2 \rangle = 1 \, , \quad i = 1,2 \tag{10.45a}$$

$$\langle u_1 \Delta_1 \rangle = \langle u_2 \Delta_2 \rangle = 0 \tag{10.45b}$$

$$\langle u_1 \Delta_2 \rangle = -X e^{-R^2/2} \tag{10.45c}$$

$$\langle u_2 \Delta_1 \rangle = Xe^{-R^2/2} \qquad\qquad (10.45d)$$

$$\langle \Delta_1 \Delta_2 \rangle = (1 - X^2)e^{-R^2/2} \qquad\qquad (10.45e)$$

$$\langle u_1 u_2 \rangle = e^{-R^2/2} \qquad\qquad (10.45f)$$

where X is the x-component of $\vec{R} = \dfrac{(\vec{r}_1 - \vec{r}_2)}{\bar{R}}$.

A relatively simple way to evaluate the average in (10.42) is to introduce a set of orthogonal variables using the Gram-Schmidt procedure. These variables $(\bar{u}_1, \bar{u}_2, \bar{\Delta}_1, \bar{\Delta}_2)$ are given by

$$\bar{u}_1 = u_1 \qquad\qquad (10.46a)$$

$$\bar{u}_2 = u_2 - \langle u_2 u_1 \rangle u_1 \qquad\qquad (10.46b)$$

$$\bar{\Delta}_1 = \Delta_1 - \frac{\langle \Delta_1 \bar{u}_2 \rangle}{\langle \bar{u}_2^2 \rangle} \bar{u}_2 \qquad\qquad (10.46c)$$

$$\bar{\Delta}_2 = \Delta_2 - \langle \Delta_2 u_1 \rangle u_1 - \frac{\langle \Delta_2 \bar{u}_2 \rangle}{\langle \bar{u}_2^2 \rangle} \bar{u}_2 - \frac{\langle \Delta_2 \bar{\Delta}_1 \rangle}{\langle \bar{\Delta}_1^2 \rangle} \bar{\Delta}_1 \qquad\qquad (10.46d)$$

and have been constructed to satisfy

$$\langle u_1 \bar{u}_2 \rangle = \langle u_1 \bar{\Delta}_1 \rangle = \langle u_1 \bar{\Delta}_2 \rangle = \langle \bar{u}_2 \bar{\Delta}_1 \rangle = \langle \bar{u}_2 \bar{\Delta}_2 \rangle = \langle \bar{\Delta}_1 \bar{\Delta}_2 \rangle = 0. \qquad (10.47)$$

Thus S_{xx} can be written as

$$S_{xx} = \frac{1}{\bar{R}^2} \langle \{ \bar{\Delta}_1 + \frac{\langle \Delta_1 \bar{u}_2 \rangle}{\langle \bar{u}_2^2 \rangle} \bar{u}_2 \} \{ \bar{\Delta}_2 + \langle \Delta_2 u_1 \rangle u_1 + \frac{\langle \Delta_2 \bar{u}_2 \rangle}{\langle \bar{u}_2^2 \rangle} \bar{u}_2 +$$

$$+ \frac{\langle \Delta_2 \bar{\Delta}_1 \rangle}{\langle \bar{\Delta}_1^2 \rangle} \bar{\Delta}_1 \} \delta(u_1) \delta(\bar{u}_2 + \langle u_2 u_1 \rangle u_1) \rangle . \qquad\qquad (10.48)$$

Due to the factor $\delta(u_1)$ in (10.48) the delta function $\delta(\bar{u}_2 + \langle u_2 u_1 \rangle u_1)$ in this expression can be replaced by $\delta(\bar{u}_2)$. Similarly, due to this factor $\delta(\bar{u}_2)$, (10.48) can be reduced to

$$S_{xx} = \frac{1}{\bar{R}^2} \langle \bar{\Delta}_1 \{ \bar{\Delta}_2 + \frac{\langle \Delta_2 \bar{\Delta}_1 \rangle}{\langle \bar{\Delta}_1^2 \rangle} \bar{\Delta}_1 \} \delta(u_1) \delta(\bar{u}_2) \rangle . \tag{10.49}$$

But by construction $\bar{\Delta}_1, \bar{\Delta}_2, u_1$ and $\bar{u}_2$ are uncorrelated, so (10.49) reduces to

$$S_{xx} = \frac{1}{\bar{R}^2} \langle \bar{\Delta}_1 (\bar{\Delta}_2 + \frac{\langle \Delta_2 \bar{\Delta}_1 \rangle}{\langle \bar{\Delta}_1^2 \rangle} \bar{\Delta}_1) \rangle \times \langle \delta(u_1) \rangle \times \langle \delta(\bar{u}_2) \rangle$$

i.e.

$$S_{xx} = \frac{1}{\bar{R}^2} \langle \Delta_2 \bar{\Delta}_1 \rangle \times \langle \delta(u_1) \rangle \times \langle \delta(\bar{u}_2) \rangle \tag{10.50}$$

since $\langle \bar{\Delta}_1 \bar{\Delta}_2 \rangle = 0$. It is straightforward to evaluate $\langle \Delta_2 \bar{\Delta}_1 \rangle$, since (10.46c),

$$\langle \Delta_2 \bar{\Delta}_1 \rangle = \langle \Delta_1 \Delta_2 \rangle - \frac{\langle \Delta_1 \bar{u}_2 \rangle \langle \bar{u}_2 \Delta_2 \rangle}{\langle \bar{u}_2^2 \rangle} . \tag{10.51}$$

From (10.46b)

$$\langle \bar{u}_2^2 \rangle = \langle u_2^2 \rangle - \langle u_2 u_1 \rangle^2 = 1 - e^{-R^2} \tag{10.52}$$

where we have used (10.45a) and (10.45f). Also, from (10.46b), (10.45b) and (10.45c)

$$\langle \Delta_1 \bar{u}_2 \rangle = \langle \Delta_1 u_2 \rangle - 0 , \quad \langle \Delta_1 \bar{u}_2 \rangle = Xe^{-R^2/2} \tag{10.53}$$

and likewise

$$\langle \bar{u}_2 \Delta_2 \rangle = 0 - \langle u_2 u_1 \rangle \langle u_1 \Delta_2 \rangle , \quad \langle \bar{u}_2 \Delta_2 \rangle = Xe^{-R^2} . \tag{10.54}$$

Thus we obtain from (10.51)-(10.54) and (10.45e)

$$<\Delta_2 \bar{\Delta}_1> = e^{-R^2/2} - x^2 \frac{e^{R^2/2}}{e^{R^2}-1} \quad . \tag{10.55}$$

To finish the calculation of S_{xx} in (10.50) we need to evaluate $<\delta(u_i)>$, $i = 1,2$.

This can be done using the Gaussian representation (10.28) for the delta function (as in the calculation of $A(t)$) to yield

$$<\delta(u_1)> = \frac{1}{\sqrt{(2\pi<u_1^2>)}} = \frac{1}{\sqrt{2\pi}} \tag{10.56}$$

and

$$<\delta(\bar{u}_2)> = \frac{1}{\sqrt{(2\pi<\bar{u}_2^2>)}} = \frac{1}{\sqrt{2\pi}(1-e^{-R^2})^{1/2}} \tag{10.57}$$

using (10.45a) and (10.52) respectively. Alternatively, we could write

$$<\delta(\bar{u}_2)> = \frac{e^{R^2/2}}{\sqrt{2\pi}(e^{R^2}-1)^{1/2}} \quad . \tag{10.58}$$

Finally (!) we obtain from (10.50), (10.55), (10.56) and (10.58)

$$S_{xx} = \frac{1}{2\pi\bar{R}^2} \left\{ \frac{1}{(e^{R^2}-1)^{1/2}} - \frac{x^2 e^{R^2}}{(e^{R^2}-1)^{3/2}} \right\} \quad . \tag{10.59}$$

Similar results hold for $S_{yy}, \ldots$ etc. Thus in d-dimensions

$$<\nabla_1 u(\vec{r}_1,t) \cdot \nabla_2 u(\vec{r}_2,t) \delta(u(\vec{r}_1,t)) \delta(u(\vec{r}_2,t))> =$$

$$= \frac{1}{2\pi\bar{R}^2} \left\{ \frac{d}{(e^{R^2}-1)^{1/2}} - \frac{R^2 e^{R^2}}{(e^{R^2}-1)^{3/2}} \right\} \quad . \tag{10.60}$$

This can be rewritten as

$$\langle \nabla_1 u(\vec{r}_1,t) \nabla_2 u(\vec{r}_2,t) \delta(u(\vec{r}_1,t)) \delta(u(\vec{r}_2,t)) \rangle =$$

$$= \frac{1}{2\pi\bar{R}^2} \frac{1}{R^{d-1}} \frac{\partial}{\partial R} \left[\frac{R^d}{(e^{R^2}-1)^{1/2}} \right] \tag{10.61}$$

using an obvious identity. Upon substituting (10.61) in (10.11) we obtain

$$S(\vec{k},t) = \left(\frac{\Delta m_o}{k}\right)^2 \frac{1}{2\pi\bar{R}^2} \int \frac{d^d(\vec{r}_1-\vec{r}_2)}{R^{d-1}} e^{i\vec{k}\cdot(\vec{r}_1-\vec{r}_2)} \cdot$$

$$\left(\frac{\partial}{\partial R} \frac{R^d}{(e^{R^2}-1)^{1/2}}\right) \cdot \tag{10.62}$$

We can write (10.62) as (with $(\vec{r}_1-\vec{r}_2) = \bar{R}\cdot\vec{R}$)

$$S(\vec{k},t) = \left(\frac{\Delta m_o}{x}\right)^2 \frac{\bar{R}^d}{2\pi} \int_0^\infty dR \left[\frac{\partial}{\partial R} \frac{R^d}{(e^{R^2}-1)^{1/2}} \int d\Omega_d \cdot e^{i\vec{x}\cdot\vec{R}} \right. \tag{10.63}$$

where

$$\vec{x} \equiv \bar{R}\,\vec{k} \tag{10.64}$$

and we have written $d^d\vec{R} = R^{d-1}dRd\Omega_d$, where $d\Omega_d$ is the hyperspherical angular integrand in d-dimensions. Upon integrating by parts, noting that the non-zero contribution of $e^{i\vec{x}\cdot\vec{R}}$ is $\cos(\vec{x}\cdot\vec{R})$, we find

$$S(\vec{k},t) = \left(\frac{\Delta m_o}{x}\right)^2 \frac{\bar{R}^d}{2} \int_0^\infty dR \frac{R^d}{(e^{R^2}-1)^{1/2}} \cdot$$

$$\left\{ -\frac{\partial}{\partial(xR)} \int d\Omega_d \cos(xR\hat{x}\cdot\hat{R}) \right\} \cdot \tag{10.65}$$

To proceed further we must specify d. We consider the case d = 3, for which

$$I_d \equiv -\frac{\partial}{\partial(xR)} \int d\Omega_d \cos(xR\cos\theta), \quad \theta = \cos^{-1}(\hat{x}\cdot\hat{R}) , \tag{10.66}$$

becomes

$$I_3 = -2\pi \frac{\partial}{\partial(xR)} \int_0^\pi d\theta \sin\theta \cos(xR\cos\theta) = -2\pi \frac{\partial}{\partial(xR)} j_0(xR)$$

where $j_0(x)$ is a spherical Bessel function $^{*)}$. Finally since $-\frac{\partial}{\partial x} j_0(x) = j_1(x) = \sqrt{\frac{\pi}{2x}} J_{3/2}(x)$, where $J_{3/2}$ is a Bessel function of first kind, one has :

$$I_3 = 4\pi \sqrt{\frac{\pi}{2xR}} J_{3/2}(xR) . \qquad (10.67)$$

Thus, for $d = 3$

$$S(\vec{k},t) = \frac{(\Delta m_0)^2}{x} \sqrt{2\pi} \bar{R}^3 \int_0^\infty dR \frac{R^3}{(e^{R^2}-1)^{1/2}} (xR)^{-1/2} J_{3/2}(xR) . \qquad (10.68)$$

For $d = 2$, one finds

$$S(\vec{k},t) = \frac{(\Delta m_0)^2}{x} \frac{\bar{R}^2}{2} \int_0^\infty dR \frac{R^2}{(e^{R^2}-1)^{1/2}} J_1(xR) . \qquad (10.69)$$

It has become customary to use a normalized structure function

$$\bar{S}(\vec{k},t) \quad \frac{S(\vec{k},t)}{\int \frac{d^d\vec{k}}{(2\pi)^d} S(\vec{k},t)} . \qquad (10.70)$$

From (10.70), (10.68) and (10.69) one obtains the scaling form (10.1) with

$$F(x) = \frac{2}{\pi} \frac{(2\pi)^{d/2}}{x} \int_0^\infty dR \ R^d [e^{R^2}-1]^{-1/2} \cdot (xR)^{1-d/2} J_{\frac{d}{2}}(xR) \qquad (10.71)$$

where $x = k\bar{R}(t)$. The agreement between (10.1) (with $F(x)$ given by (10.71)) and the results for $S(\vec{k},t)$ obtained from Monte Carlo studies of quenched kinetic Ising antiferromagnets for $d = 2$ and $d = 3$ is quite encouraging. Extensions of this work to binary alloys and binary

---

$^{*)}$ See G. Arfken, "Mathematical methods for physicists", Academic Press, New York (1970), p. 490 and 522.

fluids would be most useful.

Finally, we note that the linear and Langer, Bar-on, Miller theories discussed in Chapter 7 deal primarily with the early stages of spinodal decomposition, before the interfaces are so well established. The work dealing with the motion of random interfaces as discussed in Chapters 9 and 10 is more appropriate for the later stages of growth.

REFERENCES - Chapter 10.

[1] J.D. Gunton, M. San Miguel and P.S. Sahni, to be published in
    Vol. 8, "Phase Transitions and Critical Phenomena", Academic Press
    (London), edited by C. Domb and J.L. Lebowitz (1983).

[2] T. Ohta, D. Jasnow and K. Kawasaki, Phys. Rev. Lett. $\underline{49}$, 1223
    (1982).

[3] K. Kawasaki and T. Ohta, Prog. Theor. Phys. $\underline{62}$, 147 (1982); $\underline{68}$,
    129 (1982).

[4] S.M. Allen and J.W. Cahn, Acta Metall. $\underline{27}$, 1085 (1979).

GENERAL REFERENCES

The following papers involve Monte Carlo studies of the kinetic antiferromagnetic Ising model, with which the theory of this chapter is
concerned :

[G1] M.K. Phani, J.L. Lebowitz, M.H. Kalos and O. Penrose, Phys. Rev.
    Lett. $\underline{45}$, 366 (1980).

[G2] P.S. Sahni, G. Dee, J.D. Gunton, M. Phani, J.L. Lebowitz and
    M. Kalos, Phys. Rev. $\underline{B24}$, 410 (1981).

[G3] P.S. Sahni, G.S. Grest and S.A. Safran, Phys. Rev. Lett. $\underline{50}$, 60
    (1983).

[G4] K. Kaski, M.C. Yalabik, P.S. Sahni and J.D. Gunton, preprint.

[G5] An interesting Monte Carlo study of the quenched q-state Potts
    model has been given by P.S. Sahni, G.S. Grest, M.P. Anderson and
    D.J. Srolowitz, Phys. Rev. Lett. $\underline{50}$, 263 (1983).

[G6] An extremely interesting discussion of interface dynamics in
    fluids has been given by K. Kawasaki and T. Ohta in a paper pre-
    sented at the Conference on Nonlinear Fluid Behavior, Boulder,
    Colorado (June 7-11, 1982).

Applied Physics B

Photophysics and Laser Chemistry

Fields and Editors:

Laser Physics and Spectroscopy

High-Resolution Laser Spectroscopy:
V.P.Chebotayev, Novosibirsk
Laser Spectroscopy: **T.W.Hänsch,** Stanford U.
Quantum Electronics: **A.Javan,** MIT
Ultrafast Phenomena: **W.Kaiser,** TU München
Laser Physics and Applications:
H.Walther, U.München

Chemistry with Lasers

Chemical Dynamics and Structure: **K.L.Kompa,** MPI
Garching
Laser-Induced Processes: **V.S.Letokhov,** Moscow
Dye Laser and Photophysical Chemistry:
F.P.Schäfer, MPI Göttingen
Laser Chemistry: **R.N.Zare,** Stanford U.

Photophysics

Optics: **W.T.Welford,** Imperial College
Nonlinear Optics and Nonlinear Spectroscopy:
T.Yajima, Tokyo U.

Editor: **H.K.V.Lotsch,**
Springer-Verlag, P.O.Box 105280,
D-6900 Heidelberg 1, Federal Republic of Germany

Special Features:
● rapid publication (3–4 months)
● no page charges for concise reports
● 50 offprints free of charge

Springer-Verlag
Berlin
Heidelberg
New York
Tokyo

Subscription information and/or **sample** copies are
available from your bookseller or directly from
Springer-Verlag, Journal Promotion Dept.,
P.O.Box 105280, D-6900 Heidelberg, FRG

Lecture Notes in Physics

Selected Issues from

Lecture Notes in Mathematics